JOHN ARTHUR

The Discovery of DNA

Unraveling Life's Blueprint

First edition

This book was professionally typeset on Reedsy.
Find out more at reedsy.com

Contents

Preface

This book tells the story of one of humanity's greatest intellectual achievements—the discovery of DNA and the revelation of life's molecular foundations. It is a story that spans less than two centuries yet encompasses a transformation in human understanding as profound as any in history. From Gregor Mendel counting pea plants in a monastery garden to scientists wielding CRISPR gene editing tools that can rewrite the code of life itself, this journey has taken us from complete ignorance about heredity to unprecedented power over the genetic programs that make us who we are.

I wrote this book because the story of DNA deserves to be told in full not just as a sequence of scientific discoveries, but as a human drama involving brilliant insights and stubborn mistakes, spectacular successes and tragic oversights, proper recognition and shameful neglect. The scientists who unraveled DNA's secrets were not infallible heroes working in isolation but real people with ambitions, biases, and flaws, collaborating and competing in ways that were sometimes noble and sometimes ethically questionable.

The discovery of DNA matters not only for its scientific significance but for its continuing impact on our lives. DNA technology shapes modern medicine, agriculture, forensics, and our understanding of human history and evolution. It raises profound questions about human nature, about our relationship with the natural world, and about the proper limits of human intervention in the biological realm. These are not merely academic questions but urgent practical and ethical challenges that affect us all.

This book is intended for anyone curious about how science works, how we came to understand heredity at the molecular level, and what this understanding means for our future. I have tried to make the science accessible without oversimplifying it, to tell engaging stories without sacrificing accuracy, and to explore ethical dimensions without imposing simplistic answers. The narrative is structured chronologically but also thematically, following the major questions that drove research forward: What are genes made of? Where are they located? How are they structured? How do they work? How can we read and manipulate them?

Each chapter can stand alone, but together they build a comprehensive picture of how DNA science emerged from earlier genetics, chemistry, and cell biology, then revolutionized our understanding of life and spawned new technologies that continue to accelerate in power and scope. The appendices provide reference materials for readers who want to delve deeper into specific experiments, explore the chemistry of DNA in detail, or learn more about the scientists whose work made this revolution possible.

Some readers will approach this book with extensive scientific background; others will come with little formal training in biology or chemistry. I have tried to serve both audiences by explaining fundamental concepts clearly while not dwelling on basics unnecessarily. Technical terms are defined when first introduced and collected in a glossary for easy reference. The narrative assumes curiosity and intelligence but not specialized knowledge.

A note on perspective: this book presents the scientific consensus as it stands in 2026, acknowledging that our understanding continues to evolve. What we know about DNA today exceeds anything imaginable when the double helix was first described in 1953, yet mysteries remain and new questions constantly emerge. Science is not a collection of immutable facts but an ongoing process of discovery, correction, and refinement. The story of DNA reminds us both of how much we have learned and how much we have yet to understand.

The power that DNA science has given us to read genomes, to edit genes, to modify organisms, potentially to redesign ourselves brings with it profound responsibilities. How we use this power will shape the future of medicine, agriculture, conservation, and perhaps human evolution itself. My hope is that this book will inform those choices by helping readers understand where this power came from, what it can and cannot do, and what values should guide its application.

Finally, a word about sources: this book draws on scientific papers, historical accounts, biographies, interviews, and the extensive secondary literature on molecular biology's history. Where controversies exist and the DNA story has several I have tried to present multiple perspectives fairly while acknowledging my own interpretations. The further reading list in Appendix E provides pathways for readers to explore topics in greater depth and form their own conclusions.

The story of DNA is ultimately a story about human curiosity, ingenuity, and the relentless drive to understand the natural world. It is a story of how science progresses through the contributions of many minds across generations, building knowledge piece by piece until suddenly a complete picture emerges. And it is an ongoing story, with the most consequential chapters perhaps yet to be written. I invite you to join this journey of discovery, to marvel at what has been achieved, and to think carefully about what should come next.

Acknowledgments

This book would not have been possible without the work of countless scientists, historians, and writers who have documented and analyzed the discovery of DNA over the past century and a half. I am deeply indebted to the primary sources the original research papers, laboratory notebooks, and correspondence that preserve the authentic voices of discovery and to the many excellent historical accounts and biographies that have interpreted and contextualized these discoveries.

I am particularly grateful to the historians of science whose meticulous research has illuminated the human dimensions of DNA's discovery: Horace Freehand Judson's "The Eighth Day of Creation" remains an indispensable resource, drawing on extensive interviews with the pioneers of molecular biology. Brenda Maddox's biography of Rosalind Franklin helped restore a more accurate and balanced understanding of her crucial contributions. Matthew Cobb's work on the genetic code filled in important gaps in the standard narrative. These and many other scholars have laid the groundwork that made this synthesis possible.

I owe thanks to the scientists who have taken time to explain complex concepts, both through their published works and in the many excellent educational resources now available. The clarity with which researchers like Francis Crick, Sydney Brenner, and later generations of molecular biologists have communicated their work to broader audiences has been invaluable. The tradition of scientists writing for general readers from James Watson's controversial "The Double Helix" to Siddhartha Mukherjee's "The Gene" demonstrates that rigorous science and engaging narrative need not be incompatible.

Several institutions deserve recognition for preserving and making accessible the historical record of molecular biology. The archives at Cold Spring Harbor Laboratory, the Welcome Library, the Rockefeller Archive Center, and numerous university special collections have maintained the papers, photographs, and artifacts that allow us to reconstruct this history. The digitization efforts that have made many original papers freely available online have democratized access to scientific history in ways previous generations could only dream of.

I am grateful to the many educators and communicators who have created resources explaining DNA science to diverse audiences from the DNA Learning Center's innovative educational programs to the countless YouTube videos, podcasts, and blog posts that make molecular biology accessible. These efforts have helped build scientific literacy and have informed my own approach to explanation and storytelling.

Special acknowledgment must go to those who have worked to highlight the contributions of overlooked scientists, particularly women and minorities whose work was often minimized or ignored in traditional histories. The efforts to properly recognize Rosalind Franklin, Martha Chase, and many others have enriched our understanding of how science actually happens and have helped correct historical injustices. These corrective histories remind us that scientific progress depends on diverse contributors and that proper attribution matters both for historical accuracy and for inspiring future generations.

I am indebted to friends and colleagues who reviewed portions of the manuscript, caught errors, and offered suggestions for improvement. Any mistakes that remain are, of course, my own responsibility. The challenge of explaining complex science accurately while maintaining narrative flow requires constant balancing, and I am grateful to those who helped me navigate that balance.

My gratitude extends to the librarians, archivists, and information professionals who facilitate research by organizing and curating the ever-expanding literature of molecular biology. In an age of information overload, the skills of these professionals in helping researchers find reliable sources and distinguish quality from noise have become more valuable than ever.

I want to acknowledge the broader scientific community whose collective work over decades and centuries has built our current understanding of DNA and genetics. Science is fundamentally collaborative each discovery builds on previous work, each researcher stands on the shoulders of predecessors. The story told in this book features individual scientists and specific discoveries, but the reality is that thousands of researchers contributed to this achievement, many of whose names never appear in historical accounts. The technicians who prepared samples, the students who ran experiments, the colleagues who offered criticism and suggestions all played essential roles.

Thanks are due to my family for their patience during the long process of researching and writing this book. The hours spent reading papers, reviewing histories, and working through multiple drafts required time and focus that might otherwise have been spent elsewhere. Their support and understanding made this project possible.

Finally, I want to acknowledge the scientists who are working today to advance DNA science and to address the ethical challenges it raises. Researchers developing gene therapies for previously incurable diseases, conservationists using DNA technology to protect endangered species, ethicists grappling with the implications of germ line editing, policymakers trying to establish appropriate regulations all are writing the next chapters of the DNA story. This book looks backward to understand how we arrived at our current capabilities, but the more important question is where we go from here. That future will be shaped by choices not yet made, by science not yet done, by people not yet born. My hope is that understanding the past can inform wisdom about the future.

To all who have contributed to our understanding of DNA from Gregor Mendel in his monastery garden to contemporary researchers in laboratories around the world thank you. This book is a tribute to your curiosity, dedication, and genius.

Introduction

The Most Important Molecule

There is a molecule inside your cells right now, coiled and folded into a space almost impossibly small, that contains the complete instructions for building you. This molecule DNA, or deoxyribonucleic acid carries the genetic program that made you human rather than any other species, that gave you your particular combination of traits inherited from your ancestors, and that orchestrates the mind-boggling complexity of your body's development and function. Every protein your body makes, every cell type that differentiates during development, every trait you inherited from your parents all trace back to information encoded in DNA's elegant structure.

DNA is, without exaggeration, the most important molecule for life as we know it. Every organism on Earth from bacteria invisible to the naked eye to blue whales weighing hundreds of tons, from single-celled yeast to redwood trees reaching toward the sky, from viruses that blur the boundary between living and non-living to human beings contemplating their own existence uses DNA to store and transmit genetic information. This universality is not coincidental. It points to a single origin for all terrestrial life, a last universal common ancestor from which every living thing descended through nearly four billion years of evolution.

Yet for most of human history, DNA was completely unknown. People observed that children resembled their parents, that traits ran in families, that selective breeding could change animal and plant characteristics over generations. But the mechanism underlying these phenomena remained utterly mysterious. What actually passed from parent to offspring? Where in the body did hereditary information reside? How was it copied and transmitted? These questions seemed unanswerable, perhaps fundamentally beyond human comprehension.

The journey from complete ignorance to detailed molecular understanding took less than two centuries a blink in the span of human history, yet a period of transformation as profound as any our species has experienced. This book tells the story of that journey: how scientists pieced together the puzzle of heredity, discovered DNA, determined its structure, deciphered how it works, and ultimately learned to read and rewrite the genetic code itself.

Why This Story Matters

The discovery of DNA ranks among humanity's supreme intellectual achievements, comparable to understanding gravity, evolution, or atomic structure. But unlike some great scientific discoveries that remain largely abstract for most people, DNA science has immediate practical implications that touch nearly every aspect of modern life.

Medicine has been transformed by DNA knowledge. We now understand disease at the molecular level, recognizing that many conditions result from specific DNA mutations. We can test for genetic diseases, predict disease risks based on DNA sequences, and increasingly, treat or even cure genetic disorders by correcting the underlying DNA defects. Cancer perhaps the most feared disease is fundamentally a disease of DNA, caused by accumulated mutations that cause cells to grow out of control. Understanding this has revolutionized cancer treatment, enabling therapies targeted to specific genetic profiles.

Agriculture relies heavily on DNA technology. The crops that feed humanity have been genetically modified to resist pests, tolerate herbicides, withstand drought, and provide enhanced nutrition. Whether you support or oppose genetically modified organisms, they now constitute a large fraction of global agriculture, and DNA science will likely become even more central to food production as climate change and population growth strain agricultural systems.

Forensics and criminal justice have been revolutionized by DNA fingerprinting. Criminals are identified and convicted based on DNA evidence left at crime scenes. Innocent people wrongly convicted have been exonerated when DNA testing excluded them as perpetrators. Paternity disputes are resolved, disaster victims are identified, and missing persons are found through DNA analysis. The power to identify individuals from biological traces has fundamentally changed law enforcement.

Our understanding of human history and evolution has been rewritten by DNA analysis. We have sequenced the genomes of extinct human species like Neanderthals, revealing that they interbred with our ancestors and that most people of European and Asian descent carry Neanderthal DNA. We have traced human migrations across continents, reconstructed the domestication of crops and animals, and settled historical questions about relationships and identities through DNA evidence. The past, it turns out, left molecular traces that we can now read.

Beyond these specific applications, DNA science has transformed how we think about life, identity, and our place in nature. We know now that we are molecular beings, that our consciousness and personality emerge from genetic programs interacting with environment and experience. We understand our kinship with all living things at the molecular level. We recognize that evolution is not just a theory but an observable process written in DNA sequences. These insights change how we see ourselves and our world.

The Power and the Questions

The capabilities that DNA science has given us are extraordinary and continue to expand. We can now:

Read entire genomes, determining the complete genetic instructions for any organism

Edit genes with precision using tools like CRISPR, correcting mutations or introducing deliberate changes

Create organisms with novel combinations of genes from different species

Synthesize DNA sequences from scratch, potentially creating entirely new organisms

Clone organisms, making genetic copies from adult cells

Modify embryos, making changes that would be inherited by future generations

These capabilities would have seemed like science fiction to the scientists who discovered DNA's structure in 1953. Yet they are realities in 2026, routinely practiced in laboratories worldwide and increasingly moving from research to practical application.

But power brings responsibility, and the power to manipulate DNA raises profound questions:

Medical Ethics: Should we edit human embryos to prevent genetic diseases? What about enhancements making people stronger, smarter, or longer-lived? Who decides what changes are acceptable? How do we ensure equitable access rather than creating genetic haves and have-notch?

Environmental Impact: Should we release genetically modified organisms into the environment, with potentially irreversible consequences? Can we use gene drives to eliminate disease-carrying mosquitoes or invasive species, even if we can't fully predict the ecological effects?

Human Identity: What does it mean to be human in an age when we can modify our own genetic code? Does genetic engineering threaten human dignity or diversity? Where should we draw lines between therapy and enhancement?

Justice and Equity: As genetic technologies become more powerful, how do we ensure they benefit all of humanity rather than exacerbating inequalities? Who owns genetic information, and how should it be protected?

Long-term Consequences: What responsibilities do we have to future generations when we make changes that will be inherited? How do we balance potential benefits against uncertain risks?

These questions don't have simple answers. They require input from science, ethics, philosophy, law, religion, and democratic deliberation. They demand that we think carefully about what kind of future we want to create and what values should guide the application of genetic technology.

A Story of Science and Scientists

This book tells the DNA story both as a scientific narrative how we came to understand DNA's structure and function and as a human story about the scientists who made the discoveries. Science is often presented as a purely rational process where objective facts lead inevitably to correct conclusions. The reality is messier and more interesting.

The scientists who discovered DNA were brilliant but also fallible. They made intuitive leaps and careful measurements, inspired guesses and systematic experiments. They helped each other and competed fiercely. They sometimes shared data generously and sometimes appropriated others' work without permission. They received recognition and were overlooked, won prizes and died unacknowledged. Some were men, some were women. Some came from privilege, others from poverty. They worked across multiple countries and disciplines, speaking different languages but united by curiosity about nature's secrets.

Their story includes triumph and tragedy. Gregor Mendel died believing his work had failed, unaware it would revolutionize biology. Rosalind Franklin's crucial contribution was minimized and she died young, unable to share the Nobel Prize her work helped earn. Oswald Avery proved DNA was the genetic material but never received the recognition he deserved. James Watson and Francis Crick achieved fame and prizes but have been criticized for how they obtained crucial data and how they portrayed others.

Understanding this human dimension matters. It shows that science progresses not through individual genius working in isolation but through collaborative effort spanning generations, with each discovery building on previous work. It reminds us that scientists are people with the same ambitions, biases, and ethical blind spots as anyone else, and that the scientific process must include safeguards against these human failings. It demonstrates that proper credit and recognition matter, both for historical accuracy and for ensuring that all contributors regardless of gender, race, or status receive acknowledgment.

How to Read This Book

This book is organized chronologically and thematically, following the major questions that drove DNA research forward. The six parts trace the journey from classical genetics through the chemical identification of DNA to the determination of its structure and function, and finally to modern applications and future implications.

Part I sets the foundation, exploring how scientists before DNA was identified discovered the basic principles of heredity and located genes on chromosomes. Part II follows the chemical detective story of identifying DNA as the genetic material. Part III describes the dramatic race to determine DNA's structure, culminating in the famous double helix. Part IV explains how scientists proved DNA works as the genetic material, deciphering the genetic code and understanding replication. Part V covers the technological revolution that gave us the ability to read and manipulate DNA. Part VI reflects on what we've learned and what choices we face going forward.

Each chapter can be read independently, though together they build a comprehensive narrative. Appendices provide reference materials: a timeline of discoveries, detailed explanations of key experiments, the chemistry of DNA, a glossary of terms, suggestions for further reading, and biographies of major contributors.

I encourage you to engage actively with the material. Ask questions. Think about what assumptions scientists made and why. Consider how you would have approached these problems. Reflect on the ethical dimensions of genetic technology and what values should guide its use. This isn't just history it's an ongoing story that will shape our future, and understanding where we've been can help inform where we should go.

The Journey Ahead

The pages that follow will take you from a monastery garden in nineteenth-century Moravia to cutting-edge laboratories applying CRISPR gene editing. You'll meet determined monks and ambitious graduate students, meticulous experimental and bold theorists, overlooked contributors and celebrated prize winners. You'll see how obscure observations about pea plants grew into a complete molecular understanding of heredity. You'll witness fierce competition and generous collaboration, brilliant insights and stubborn errors, ethical conduct and troubling lapses.

Most importantly, you'll come to understand what DNA is, how it works, how we discovered these facts, and what this knowledge means for medicine, agriculture, conservation, human identity, and our future. You'll gain appreciation for both the power and the limits of DNA science what it can tell us and what remains mysterious, what it enables us to do and what we should or shouldn't do with that power.

The discovery of DNA is fundamentally a story about human curiosity and ingenuity our drive to understand the world and ourselves, our ability to piece together complex puzzles through careful observation and creative thinking, our capacity to build knowledge across generations. It's also a story about responsibility the obligation to use knowledge wisely, to consider consequences before acting, to ensure benefits are shared equitably, and to respect both human dignity and the natural world.

As you read, remember that this story isn't finished. Scientists today are writing new chapters, developing new technologies, making new discoveries, and grappling with new ethical challenges. The power to read and rewrite the code of life continues to grow. How we use that power whether with wisdom or recklessness, for benefit or harm, fairly or unjustly remains to be determined. Understanding the journey that brought us here is the first step toward ensuring that where we go next is guided by knowledge, ethics, and genuine concern for human welfare and the flourishing of all life.

The most important molecule awaits. Let us begin.

One

The Ancient Question

"What Makes Us Who We Are?"

The child has her mother's eyes. Everyone can see it the same deep brown, the same gentle curve at the corners, the same way they crinkle when she laughs. Her father's nose sits prominently on her face, and there's no denying the distinctive shape that has marked three generations of his family. When she walks, she carries herself with her grandmother's peculiar grace, and when she speaks, her voice holds an echo of her mother's melodic cadence. We look at this child and see a living tapestry woven from threads of her ancestors, a breathing testament to the mysterious force that connects parent to child across the infinite chain of generations.

For as long as humans have had the capacity to wonder, we have asked ourselves how this magic works. What invisible hand reaches across the divide between one generation and the next, ensuring that children resemble their parents, that siblings share certain features, that family characteristics persist through time like an unbroken song? This question has haunted humanity's greatest thinkers, driven our most ambitious farmers and breeders, and ultimately launched one of the most revolutionary discoveries in the history of science the discovery that would reveal the very molecule of life itself.

The Ancients Ponder the Mystery

Long before microscopes revealed the hidden world within our cells, long before chemistry gave us the language to speak of molecules and atoms, ancient philosophers wrestled with the enigma of inheritance. In the schools of ancient Greece, where philosophy and observation merged in humanity's first systematic attempts to understand nature, two competing visions emerged that would echo through the centuries.

Aristotle, perhaps the most influential natural philosopher of the ancient world, proposed that heredity arose from the mixing of contributions from both parents. In his view, which he laid out with characteristic confidence in the fourth century BCE, the mother provided the raw material the substance and matter that would become the child's body. The father, however, contributed something altogether different and, in Aristotle's decidedly patriarchal view, more important: the form, the vital force that would shape that matter into a new human being. Think of it as the difference between clay and the sculptor's vision. The mother provided the clay, passive and waiting, while the father's contribution was like the sculptor's hands and imagination, actively shaping and organizing that raw material into a living, breathing person.

This wasn't just philosophical speculation divorced from reality. Aristotle had observed the natural world with remarkable care, studying the development of chicken eggs and watching how seemingly formless matter gradually took on the distinct shapes and structures of a living creature. He saw this transformation and concluded that inheritance must involve some organizing principle, some force that directed development. What he couldn't see, of course, was the molecular machinery that actually makes this happen, but his instinct that there must be instructions, some kind of information being passed along, was remarkably prescient.

Competing with Aristotle's vision was an older idea, often attributed to Hippocrates and other earlier Greek thinkers, which held that heredity involved the collection of materials from all parts of both parents' bodies. Imagine tiny representatives traveling from every organ, every limb, every feature, gathering together to form the seed that would become the next generation. Your eyes contributed something to the mix, your hands added their essence, your liver and heart and brain all sent delegates to the convention that would determine your offspring's characteristics. This theory, called pan genesis, had an intuitive appeal it seemed to explain why children could inherit traits from either parent, and why they might resemble their parents in multiple, specific ways rather than just bearing a vague family resemblance.

Both theories were wrong in their particulars, of course, but both contained seeds of truth that would only be understood more than two thousand years later. Aristotle was right that development involves information and instructions, though he misunderstood where those instructions resided and how they worked. The pan genesis theorists were right that something material must pass from parent to child, though the mechanism they imagined bore no resemblance to the actual molecular processes we now understand.

The Long Sleep of Understanding

For nearly two millennia after the great Greek philosophers laid out these competing visions, humanity made remarkably little progress in understanding heredity. The question didn't disappear far from it but the tools and concepts needed to crack the mystery simply didn't exist yet. Medieval scholars, working within frameworks dominated by religious authority and ancient texts, largely accepted one or another of the classical theories without feeling much need to test them through systematic observation or experiment.

Yet even in the absence of scientific understanding, practical knowledge accumulated quietly in fields and pastures, in breeding barns and seed selection, in the careful hands of farmers who might not have been able to explain why their methods worked but knew with certainty that they did work. Shepherds understood that breeding their largest rams with their most robust ewes would tend to produce strong offspring. Grain farmers learned to save seeds from their most productive plants, gradually improving their crops over generations through patient selection. Pigeon fanciers developed exotic varieties with spectacular plumage or unusual body shapes, demonstrating that inheritance could be molded and shaped by careful breeding choices.

This practical knowledge represented a profound truth: heredity was not random or capricious but followed patterns that could be understood and even manipulated. Like begets like, the farmers would say, capturing in simple words an observation that contradicted any notion of purely mystical or arbitrary inheritance. Tall parents tend to have tall children. Fast horses sired by fast horses tend to be fast themselves. White sheep bred with white sheep produce white lambs, while black bred with black produces black.

But knowing that something happens is vastly different from understanding how or why it happens. The farmers and breeders were practicing what we might call applied heredity without having any theoretical framework to explain their successes. They were like people using computers without understanding electricity, circuits, or binary code capable of achieving remarkable results while remaining completely ignorant of the underlying mechanisms that made those results possible.

The Dawn of Scientific Inquiry

The scientific revolution that swept through Europe in the sixteenth and seventeenth centuries brought with it new ways of thinking about the natural world, new instruments for observing phenomena too small or too distant for naked human senses, and most importantly, a new conviction that nature's secrets could be unlocked through careful observation, systematic experimentation, and mathematical reasoning. Galileo pointed his telescope at the heavens and saw moons circling Jupiter, demonstrating that Earth was not the center of all motion. Newton showed that the same laws governing falling apples also governed the movements of planets, uniting terrestrial and celestial mechanics under a single mathematical framework. Chemists began to understand that matter was composed of elements that combined according to fixed proportions and rational principles.

In this heady atmosphere of discovery, scientists naturally turned their attention to the mystery of heredity with renewed vigor and improved tools. The invention of the microscope in the late sixteenth century opened up an entirely new world a universe that had always existed but had remained completely invisible to human eyes. When researchers first peered through these remarkable instruments at drops of pond water, at slices of cork, at their own blood, they discovered that life existed at scales they had never imagined. Single-celled organisms swam and fed and reproduced in every droplet of water. Plant tissues revealed intricate cellular architectures. Human blood teemed with countless cells floating in liquid plasma.

Among the most exciting early microscopic discoveries was the observation of sperm cells, first seen by Anton van Leeuwenhoek in 1677 when he examined his own semen under one of his handcrafted microscopes. Here, at last, was something material passing from father to offspring thousands upon thousands of tiny, swimming organisms, each with a round head and a long, whipping tail. Some enthusiastic microscopic even claimed they could see tiny, fully-formed people curled up inside sperm heads, waiting to be planted in the mother's womb where they would simply grow larger. This theory, called reformation, suggested that development wasn't really development at all but merely the enlargement and unfolding of structures that already existed in miniature, perhaps nested inside each other like an infinite set of Russian dolls stretching back to Adam and Eve.

The competing theory, epigenesis, held that offspring developed gradually from unformed material, with structures and organs appearing in sequence rather than all existing from the beginning in tiny form. This view seemed more consistent with actual observations of developing embryos, which began as seemingly simple, unstructured masses and only gradually took on recognizable forms. Yet it raised a troubling question: if the embryo starts formless, what directs it to take on the right form, to develop into a chicken rather than a fish, into a human rather than a pig?

Both theories stumbled over the same fundamental problem that had plagued Aristotle and Hippocrates two thousand years earlier: they lacked any understanding of how information could be stored, transmitted, and expressed in living matter. The preformationists had information in the wrong place they thought it was embodied in actual, physical structures somehow compressed to microscopic size. The epigenesists recognized that development involved a progressive building of complexity, but they had no explanation for what directed that building, what kept it on track, what ensured that generation after generation of organisms faithfully reproduced their kind.

The Question Crystallizes

By the mid-nineteenth century, the mystery of heredity had reached a critical juncture. Centuries of observation and breeding had accumulated a vast body of practical knowledge. Microscopes had revealed cells, sperm, eggs, and the beginnings of embryonic development. Yet the fundamental mechanism remained as elusive as ever. Scientists knew that children inherited traits from both parents, not just one. They knew that inherited traits could skip generations, appearing in grandchildren even when absent in parents. They knew that new varieties could be created through careful breeding. But they didn't know how any of this actually worked.

The stage was set for a revolution, though few people realized it at the time. What was needed was someone willing to approach the problem not with grand philosophical theories or purely descriptive observations, but with careful, systematic experiments someone who would count and measure and record with painstaking precision, someone who would let the numbers speak for themselves rather than trying to force observations to fit preconceived notions.

That someone was tending a garden in a monastery in central Europe, about to embark on an eight-year experimental program that would, after decades of neglect and eventual rediscovery, finally crack open the mystery that had haunted humanity since the first humans had sense enough to wonder. In that quiet monastery garden, surrounded by thousands of pea plants carefully tended and meticulously documented, Gregor Mendel was about to discover the fundamental laws of inheritance laws written not in philosophy or theology but in the language of mathematics, laws that would eventually point toward a chemical answer to the ancient question of what makes children resemble their parents.

The answer, when it finally came, would be more beautiful and more profound than anyone had imagined. It would be written in a molecule, a twisted ladder of stunning elegance, a code that all living things share, a chemical thread connecting every organism on Earth to the very first spark of life billions of years ago. But before we can understand DNA, before we can appreciate the magnificent complexity of the genetic code, we must first understand the experiments that revealed its existence, the brilliant minds that deciphered its structure, and the revolutionary implications that continue to reshape our world.

The journey begins with peas in a monastery garden. It ends with the power to read and rewrite the very instructions that make us human. This is the story of the greatest detective story ever told the hunt for the molecule of life itself.

Two

Mendel's Mathematical Garden

"The Monk Who Counted Peas"

On a warm summer morning in 1856, in the peaceful city of Brno in what is now the Czech Republic, a thirty-four-year-old Augustinian friar walked through the gardens of his monastery carrying a notebook, a magnifying glass, and a set of small paintbrushes. His name was Gregor Johann Mendel, and to the casual observer, he appeared to be doing nothing more remarkable than tending the monastery's vegetable plots. No one watching him that day could have guessed that this unassuming man, moving methodically from plant to plant, carefully examining flowers and recording observations in his meticulous handwriting, was about to revolutionize humanity's understanding of heredity and lay the foundation for the entire field of genetics.

Mendel's path to that garden had been neither straight nor easy. Born in 1822 to a peasant farming family in the village of Heinzendorf in Austrian Silesia, young Johann Mendel had shown early academic promise despite the poverty that surrounded him. His teachers recognized something special in the boy a keen mind, an unusual capacity for mathematical thinking, and a relentless curiosity about the natural world. But promise alone couldn't pay for education, and Johann's family struggled to support his studies. His younger sister even sacrificed part of her dowry to help pay his fees, a gesture he would remember with gratitude for the rest of his life.

At twenty-one, Johann made a decision that would shape his future in ways he couldn't fully anticipate. He entered the Augustinian monastery of St. Thomas in Brno, taking the name Gregor and joining a community that, remarkably for its time, valued learning and scientific inquiry as highly as prayer and contemplation. The monastery was no mere retreat from the world but rather a center of intellectual activity, where friars pursued studies in physics, mathematics, meteorology, and natural history. The abbot, Cyril Napp, was himself a man of considerable learning who encouraged scientific investigation and maintained connections with the broader European scientific community. Under Napp's leadership, the monastery had become something of an academy, where men of scholarly inclination could pursue knowledge while fulfilling their religious obligations.

Mendel tried his hand at parish work, but he was temperamentally unsuited for pastoral duties the suffering and death he witnessed as a priest ministering to the sick left him emotionally devastated. Recognizing this, Abbot Napp suggested that Mendel might be better suited to teaching, and arranged for him to attend the University of Vienna to study physics, mathematics, and natural sciences. There, between 1851 and 1853, Mendel studied under some of the most distinguished scientists of his day, including Christian Doppler of the famous Doppler effect, and absorbed the rigorous experimental methods and statistical approaches that would later prove crucial to his breakthrough in understanding heredity.

The Garden Laboratory

When Mendel returned to Brno after his university studies, he brought with him a passion for experimental science and a particular interest in the question that had bedeviled naturalists for centuries: how do traits pass from parents to offspring? The monastery's gardens provided him with the perfect laboratory. Here he had space, time, resources, and most importantly, the patient support of his religious community. Abbot Napp even allocated a dedicated plot for Mendel's experiments and exempted him from some duties to give him time for his research.

But Mendel was no naive amateur stumbling into discovery by accident. Before he ever planted his first experimental pea, he spent two years in preliminary studies, carefully selecting and testing different varieties of plants to determine which would best serve his experimental purposes. He needed a plant that could be easily grown in large numbers, that had distinct and easily observable traits, that could be cross-pollinated by human intervention but would normally self-pollinate naturally, and that would produce multiple generations quickly so he could observe inheritance patterns over time.

After extensive trials, Mendel settled on the common garden pea, Pi sum satsuma, as his experimental subject. It was a choice of remarkable insight. Peas had several invaluable characteristics that made them ideal for heredity research. They grew relatively quickly, producing a new generation in a single growing season. They had numerous varieties with clear-cut, contrasting traits that didn't blur together but remained distinct. Most crucially, peas were normally self-pollinating each flower fertilized itself unless someone intervened to prevent it which meant that Mendel could maintain pure-breeding lines that consistently produced offspring identical to the parents, and he could control exactly which plants bred with which through careful manipulation.

Mendel identified seven pairs of contrasting characteristics in his peas, each representing a simple either-or choice with no intermediate forms. The plants were either tall or short, never medium height. The seeds were either round or wrinkled in shape, never somewhere in between. Seed color was either yellow or green, pod color was green or yellow, pod shape was inflated or constricted, flower color was purple or white, and flowers appeared either along the stems or clustered at the stem tips. These clear distinctions would prove absolutely essential to his eventual breakthrough, allowing him to count and categorize offspring with precision and without ambiguity.

The Eight-Year Experiment

What Mendel undertook next would have broken the spirit of a less patient man. Between 1856 and 1863, he grew and examined nearly thirty thousand pea plants, conducting countless crosses and meticulously recording the characteristics of every single plant. He didn't just observe casually and note general patterns he counted everything. How many tall plants resulted from each cross? How many short? How many plants had purple flowers versus white? He recorded every number, every ratio, every detail in careful tables that filled page after page of his notebooks.

The work was painstaking beyond measure. Each flower that Mendel wanted to cross-pollinate had to be carefully opened before it could self-pollinate, the male parts removed with fine forceps to prevent self-fertilization, and then pollen from another plant had to be transferred to the female parts using a small brush. After pollination, Mendel had to protect the flowers to ensure no stray pollen from insects or wind could contaminate his results. When the seeds formed, he had to plant them, nurture them to maturity, observe and record their characteristics, and then begin the process again for the next generation.

Imagine the scene: Mendel in his black friar's habit, moving through rows of plants in the monastery garden, bending to examine flowers with his magnifying glass, carefully opening blossoms with steady hands, transferring pollen with delicate brushstrokes, tying protective bags over fertilized flowers, and always, always recording observations in his notebook. Day after day, week after week, year after year. His fellow friars must have wondered at his dedication to such repetitive work, but Mendel knew something they didn't he knew that truth reveals itself not in casual observation but in careful measurement, not in small samples but in large numbers, not in anecdotes but in statistics.

The Pattern Emerges

The breakthrough came gradually as Mendel analyzed the numbers accumulating in his notebooks. When he crossed pure-breeding tall plants with pure-breeding short plants, something remarkable happened. The first generation offspring what Mendel called the "first filial" generation, abbreviated F_1 were all tall. Every single one. It was as if the shortness had vanished completely, dominated and suppressed by the tallness. But Mendel didn't stop there. He allowed these F_1 plants to self-pollinate and produce a second generation, the F_2 generation.

Here the mystery deepened in the most wonderful way. In the F_2 generation, shortness reappeared, but not in random or unpredictable proportions. When Mendel counted the plants and he counted thousands of them to be sure he found that approximately three-quarters were tall and one-quarter were short. The ratio was remarkably consistent: roughly 3:1, tall to short. It wasn't exactly three to one in every batch, of course random variation saw to that but when he pooled all his data together, the ratio held with striking precision. Out of 1,064 F_2 plants in one set of crosses, he found 787 tall and 277 short, almost exactly a 3:1 ratio.

The same pattern appeared with every trait he examined. When he crossed round-seeded plants with wrinkled-seeded plants, the F_1 generation all had round seeds, and the F_2 generation showed roughly three-quarters round and one-quarter wrinkled. Yellow seeds crossed with green seeds produced all yellow in the F_1 generation, then three-quarters yellow and one-quarter green in the F_2. Purple flowers crossed with white flowers behaved the same way: all purple in F_1, then three purple to one white in F_2. The mathematical consistency was undeniable, and it pointed to something profound.

Mendel realized that inheritance could not work through any blending or mixing of parental traits, as most scientists assumed. If traits blended like paint colors, crossing tall and short plants should produce medium-height offspring, and those medium plants breeding together should produce more medium plants. Instead, the traits remained distinct and separate across generations, disappearing in one generation only to reappear unchanged in the next. The shortness that vanished in the F_1 generation returned in the F_2 precisely as short as it had been in the original parent, not diluted or altered by its time hidden from view.

The Revolutionary Insight

Sitting in his study, analyzing the tables of numbers that represented years of backbreaking work, Mendel made a conceptual leap that would eventually transform biology. He proposed that each trait was controlled by a pair of "factors" (we now call them genes, but Mendel had no name for them beyond "factors"). Each parent contributed one factor to their offspring, so every offspring received a pair one from mother, one from father. Some factors were dominant, meaning they would produce their effect even when paired with a different factor. Other factors were recessive, meaning they could be hidden by a dominant factor but would still be passed along to the next generation intact and unchanged.

The elegant simplicity of this explanation accounted for everything Mendel had observed. When a pure-breeding tall plant (having two "tall" factors) crossed with a pure-breeding short plant (having two "short" factors), the offspring all received one tall factor and one short factor. Because tall was dominant, all the F_1 plants appeared tall. But they weren't identical to their tall parent hidden within them was the short factor, waiting silently, biding its time.

When these F_1 plants self-pollinated, their factors could combine in new ways. Some offspring would receive two tall factors and be tall. Some would receive one tall and one short factor and still be tall because tall was dominant. But some approximately one-quarter of them would receive two short factors, and with no tall factor to dominate, their shortness would finally express itself. The mathematics worked out perfectly: if you consider all the possible combinations when each parent randomly contributes one of their two factors, you get a 3:1 ratio of dominant to recessive traits in the offspring. Mendel had discovered the fundamental mathematical laws of inheritance.

He went further, conducting more complex crosses that tested his model rigorously. He examined multiple traits simultaneously, crossing plants that differed in two characteristics at once seed shape and seed color, for example. His principle of independent inheritance predicted that the factors for different traits would sort independently during reproduction, producing offspring in predictable ratios: 9:3:3:1 for two traits, even more complex ratios for three or more traits. And that's exactly what his experiments revealed. The numbers from his carefully tended plants matched his mathematical predictions with stunning accuracy.

The Tragedy of Timing

On February 8 and March 8, 1865, Mendel presented his findings to the Natural History Society of Brno in two lectures that together represented one of the most important scientific communications in history. His audience of about forty local naturalists and educated citizens listened politely, but they showed little enthusiasm and apparently understood even less. The concepts Mendel presented were too new, too mathematical, too abstract for minds trained in descriptive natural history rather than experimental science. When one attendee asked questions, it became clear that he had completely missed Mendel's point, focusing on irrelevant details rather than grasping the revolutionary principles.

The following year, 1866, Mendel published his work in the Proceedings of the Natural History Society of Brno under the title "Experiments on Plant Hybridization." It was a remarkably clear and thorough paper, forty-four pages that laid out his methods, his data, and his conclusions with exemplary scientific rigor. He sent copies to prominent scientists and institutions throughout Europe, hoping to spark interest and recognition. The response was a deafening silence. No one grasped the significance of what he had discovered. No one repeated his experiments. No one built upon his findings. The paper sat in obscure journals and library shelves, gathering dust while the scientific world continued to puzzle over heredity using the same tired theories that Mendel had already demolished with his data.

Part of the problem was that Mendel was ahead of his time in multiple ways. His mathematical approach to biology was alien to most naturalists of the 1860s, who thought in terms of descriptions and classifications rather than numbers and ratios. His concept of discrete hereditary factors contradicted the prevailing belief in blending inheritance. Most fundamentally, without any knowledge of cells dividing, chromosomes separating, or genes as physical entities, scientists lacked the conceptual framework to understand what Mendel's factors actually were or where they might be located. Mendel had discovered the rules of a game without seeing the players or the board.

There was another problem too: Darwin. In 1859, just three years before Mendel began publishing his results, Charles Darwin had published "On the Origin of Species," unleashing a storm of controversy and debate that dominated biological thinking for decades. Darwin's theory of evolution by natural selection was revolutionary, controversial, and all-consuming. It demanded attention, provoked fierce arguments, and reshaped the entire landscape of biological thought. Next to this intellectual earthquake, Mendel's quiet paper about pea plants in a provincial journal seemed impossibly obscure and irrelevant. Even Darwin himself, who desperately needed a theory of inheritance to support his evolutionary ideas, never learned of Mendel's work, though a copy of the paper sat unread in Darwin's library.

In 1868, Mendel was elected abbot of his monastery, a position that brought honor and responsibility but also an end to his scientific research. Administrative duties consumed his time, and a protracted conflict with the government over taxation of religious institutions drained his energy. He continued to pursue some experiments with other plants, particularly hawk weed, but those species proved less cooperative than his beloved peas, showing patterns of inheritance complicated by factors Mendel couldn't have known about. His later work yielded no new breakthroughs, and as the years passed, even Mendel himself seemed to lose hope that his discoveries would ever be recognized.

Gregor Mendel died in 1884 at the age of sixty-one, believing himself a failure. His groundbreaking paper had been cited a mere handful of times, always in passing and never with true understanding. The new abbot, cleaning out Mendel's study, burned most of his papers and notes, unknowingly destroying irreplaceable records of one of history's greatest scientific achievements. The monastery garden where Mendel had made his discoveries was eventually paved over. It seemed that Mendel's life's work would disappear without a trace, his name forgotten, his discoveries lost forever to the progress of science.

But buried ideas, like buried seeds, sometimes wait in darkness until conditions are right for them to emerge into the light. Sixteen years after Mendel's death, in the spring of 1900, three scientists working independently in different countries would simultaneously rediscover his laws, finally giving the patient monk the recognition he deserved. The seeds he had planted, both literal and metaphorical, would finally bear fruit, and the science of genetics would burst into bloom with a speed and vigor that would have astonished even Mendel himself. His factors would be renamed genes, their physical basis would be discovered, and ultimately scientists would identify the very molecule that carried his hereditary information from generation to generation.

But that revelation still lay in the future. For now, as the nineteenth century drew to a close, scientists still didn't know what Mendel's factors actually were, where they resided in the cell, or what they were made of. The next chapter in our story would take researchers on a journey into the cell itself, where a new generation of scientists armed with improved microscopes and innovative staining techniques would discover the structures that housed Mendel's mysterious factors the threadlike chromosomes that held the key to inheritance.

Three

The Chromosomal Connection

"Watching Inheritance Under the Microscope"

While Mendel's paper gathered dust in obscure libraries across Europe, another revolution was quietly unfolding in laboratories equipped with increasingly powerful microscopes and ingenious new chemical techniques for staining cells. Scientists were discovering that the cell that fundamental unit of life that had been invisible to humanity until the invention of the microscope two centuries earlier was not the simple blob of jelly that early observers had imagined. Instead, cells revealed themselves to be fantastically complex structures containing specialized compartments, intricate machinery, and most intriguingly, a curious collection of threadlike objects that appeared and disappeared in an almost choreographed dance whenever cells divided.

The journey into this microscopic world required not just better instruments but also new ways of preparing specimens for observation. Living cells are largely transparent, their internal structures invisible even under powerful magnification. The breakthrough came with the development of synthetic dyes chemicals that would selectively bind to certain cellular components and make them visible. These staining techniques, borrowed and adapted from the thriving textile industry of the mid-nineteenth century, transformed microscopy from an exercise in squinting at barely visible shadows into a revelation of cellular architecture in vivid color.

The Dance of Division

German biologist Walter Fleming stands as one of the heroes of this microscopic exploration. Working in the 1870s and early 1880s at the University of Kiel, Fleming became fascinated by what happened inside cells during division, the process by which one cell becomes two. Previous observers had noticed that cells divided, but the details remained frustratingly murky. Fleming, armed with improved microscopes and a particularly effective staining technique using a dye called aniline, saw what no one had clearly seen before.

Inside the cell's nucleus that spherical structure at the cell's center Fleming observed threadlike structures that absorbed the stain intensely, becoming brightly colored and clearly visible. He called these structures "chromatin" (from the Greek word for color) because of their affinity for dyes, and he watched with growing excitement as they went through an elaborate transformation during cell division. The chromatin would condense and organize itself into distinct, individual threads, which Fleming named "chromosomes," meaning literally "colored bodies."

What Fleming witnessed and meticulously documented over hundreds of hours of observation was nothing short of spectacular. As a cell prepared to divide, its chromosomes would become visible, then line up in an orderly fashion across the cell's middle, as if marshaling themselves for some crucial maneuver. Then, in a movement of remarkable precision, each chromosome would split lengthwise down its center, and the two halves would migrate to opposite ends of the cell, pulled by invisible forces that Fleming could observe in their effects but could not yet explain. Finally, the cell itself would pinch in two, creating two daughter cells, each containing a complete set of chromosomes identical to those in the original parent cell.

Fleming called this process "mitosis," from the Greek word for thread, and he recognized immediately that it was far too elaborate and precise to be a random or unimportant event. The cell went to extraordinary lengths to ensure that each daughter cell received exactly the same chromosomal content as its parent. This suggested powerfully that chromosomes must contain something vitally important, something that absolutely had to be duplicated and distributed equally to each new cell. But what that something might be remained a tantalizing mystery.

A Second Kind of Division

As researchers studied cell division in different tissues and organisms, they discovered something even more remarkable. In the specialized cells that would become sperm and eggs the gametes that would unite in fertilization to create a new organism cell division followed a different pattern. This special form of division, eventually named meiosis (from the Greek word for diminution or reduction), involved not one but two successive divisions, and it produced something quite unexpected: cells containing exactly half the number of chromosomes as normal body cells.

The implications were staggering once scientists recognized them. If sperm and eggs each contained half the normal chromosome number, then when they fused during fertilization, the resulting offspring would have the full complement restored half from the mother, half from the father. This wasn't just an elegant solution to what might otherwise be an exponentially expanding chromosome problem (if gametes had the full number, offspring would have double, their offspring quadruple, and so on). It was also a mechanism by which offspring could receive equal contributions from both parents.

Belgian zoologist Édouard van Beneden made crucial observations in this area while studying parasitic worms in the 1880s. Working with the intestinal worm Ascaris, which had the convenient feature of possessing only a few very large chromosomes that were easy to see and count, van Beneden demonstrated definitively that the chromosomes from sperm and egg remained distinct during fertilization. He could actually see the two sets come together in the newly fertilized egg, neither dominating nor destroying the other, but coexisting side by side. Later, when this fertilized egg began to divide, its chromosomes would duplicate and distribute themselves according to the mitotic pattern Flemming had described, ensuring that every cell in the developing worm carried chromosomes derived from both parents.

The Light Dawns

By the turn of the twentieth century, the pieces of an enormous puzzle were scattered across laboratory benches throughout Europe and America, waiting for someone to see how they fit together. In 1900, that moment finally arrived, triggered by a botanical curiosity that sent three scientists rushing to dusty library shelves where a forgotten paper by an obscure monk had been waiting for thirty-four years.

Hugo de Vries in the Netherlands, Carl Correns in Germany, and Erich von Tschermak in Austria were all conducting plant breeding experiments, and all three independently rediscovered the same mathematical ratios and principles that Mendel had discovered decades earlier. Each of them, in that deflating moment that scientists both dread and embrace, found that they had been scooped by a man who had been dead for sixteen years. To their credit, all three acknowledged Mendel's priority, and almost overnight, Mendel's work vaulted from obscurity to fame. The science of genetics a name coined in 1906 by the English biologist William Bateson was born, and Mendel's factors finally received the recognition they deserved.

But it was the crucial insight that came shortly afterward, connecting Mendel's abstract factors with the very real, very visible chromosomes that microscopic had been studying, that truly revolutionized biology. Two scientists working independently on opposite sides of the Atlantic made this connection almost simultaneously in 1902, though each approached it from a slightly different angle.

Walter Sutton was a young American graduate student at Columbia University, working with grasshopper chromosomes and bringing to his observations a strong background in both cell biology and Mendelian genetics, which was now spreading rapidly through the scientific community. Sutton noticed something that seemed too perfect to be coincidental: chromosomes behaved exactly as Mendel's factors should behave. Body cells had chromosomes in pairs, just as Mendel had proposed that organisms had factors in pairs. During the formation of gametes through meiosis, these chromosome pairs separated, with each gamete receiving just one chromosome from each pair precisely parallel to Mendel's principle that each parent contributed one factor to offspring. At fertilization, chromosomes from sperm and egg combined to restore the paired condition exactly as Mendel's factors combined in offspring.

Theodor Boveri, an established German biologist working with sea urchin eggs, came to similar conclusions through elegant experiments that demonstrated the importance of having the right chromosomes in the right numbers. Boveri showed that normal development required a complete set of chromosomes, and that missing or extra chromosomes led to abnormal development or death. This proved that chromosomes weren't just structures going through pointless choreography they carried essential information needed for proper development and function.

Together, Sutton and Boveri proposed what came to be known as the chromosomal theory of inheritance: Mendel's factors now increasingly called genes were located on chromosomes. The correlation was too perfect, the parallels too exact, to be mere coincidence. Chromosomes were the physical structures that carried hereditary information from one generation to the next. The behavior of chromosomes during cell division and gamete formation explained the behavior of genes during inheritance. The mystery of where heredity resided in the cell had been solved.

Proof Through Flies

The chromosomal theory was elegant and persuasive, but science demands more than elegant theories. It demands proof, experimental evidence that goes beyond correlation to demonstrate causation. That proof would come from one of the most unlikely experimental organisms imaginable: the common fruit fly, Drosophila melanogaster, that tiny pest that appears seemingly from nowhere whenever fruit begins to rot.

Thomas Hunt Morgan, a skeptical and rigorous experimental biologist at Columbia University, initially doubted both Mendelian genetics and the chromosomal theory. He thought the enthusiasm for Mendel's rediscovered laws was overblown, and he suspected that heredity would prove far more complex and messy than the neat mathematical ratios suggested. Determined to find the exceptions and complications that would temper this genetic enthusiasm, Morgan began breeding fruit flies in 1908, filling milk bottles with flies and examining their offspring under microscopes.

Fruit flies had several advantages as experimental organisms that would eventually make them the laboratory workhorses of genetics. They were tiny, requiring minimal space and resources. They bred prodigiously, producing a new generation every two weeks. They had only four pairs of chromosomes, making genetic analysis simpler. And most importantly, they occasionally produced mutants flies with characteristics different from the normal type that could be used to trace inheritance patterns.

In May 1910, after two years of looking at countless normal red-eyed flies, Morgan noticed a male with white eyes peering at him from one of his bottles. This single mutant fly would change everything. Morgan carefully bred the white-eyed male with normal red-eyed females, and the results were striking. All the offspring had red eyes, suggesting that red was dominant over white, just as Mendel's principles would predict. But when Morgan bred these red-eyed offspring together, something remarkable emerged in the next generation: white-eyed flies reappeared, but they were all males. The white-eye trait seemed somehow linked to being male.

Morgan knew that sex itself was determined by chromosomes. Female flies had two X chromosomes, while males had one X and one smaller Y chromosome. If the gene for eye color was located on the X chromosome, everything made sense. Males, with only one X chromosome, would display whatever eye-color gene that X carried red or white. Females, with two X chromosomes, would need two copies of the white gene to have white eyes; one copy of the red gene would dominate. This explained perfectly why white eyes appeared only in males in certain crosses and demonstrated beyond doubt that genes resided on chromosomes.

Morgan's conversion from skeptic to enthusiast was complete and total. He threw himself into fruit fly genetics with passionate intensity, and his laboratory soon known affectionately as the "Fly Room" became the epicenter of genetic research. Working in a crowded, fly-filled space at Columbia University, Morgan and his brilliant students (including Alfred Stuyvesant, Calvin Bridges, and Herman Muller) mapped the locations of genes on chromosomes, demonstrated that genes could mutate and create new variations, showed that genes located on the same chromosome tended to be inherited together, and developed the mathematical tools to measure the distances between genes based on how often they were separated during meiosis.

By 1915, Morgan and his team had accumulated so much evidence supporting the chromosomal theory of inheritance that it could no longer be considered a theory at all it was simply fact. Genes were on chromosomes, arranged in linear order like beads on a string. Different versions of genes (what we now call alleles) occupied the same position on corresponding chromosomes. The behavior of chromosomes during meiosis and fertilization explained the behavior of genes during inheritance. Mendel's abstract factors had been given a physical address.

The Question Remains

Yet even with this magnificent progress, a crucial question remained unanswered. Chromosomes had been identified as the carriers of genes, but chromosomes were complex structures made of multiple components. Chemical analysis revealed that chromosomes contained two major substances: proteins and nucleic acids. Proteins were well-studied and impressively diverse, built from combinations of twenty different amino acids that could be arranged in countless ways to create molecules of enormous variety and complexity. Nucleic acids, by contrast, seemed far simpler and more monotonous, built from just four different chemical units.

Given this disparity, the conclusion seemed obvious to most scientists. Genes must be made of proteins. Proteins had the complexity, the diversity, the sophistication to encode the vast amount of information needed to direct the development and function of living organisms. Nucleic acids seemed too simple, too limited, too repetitive to be the hereditary material. They were probably just structural scaffolding, holding the chromosomes together while the proteins did the real work of heredity.

This assumption would dominate scientific thinking for decades, delaying the recognition of DNA's true role and sending researchers down blind alleys in search of the protein genes that didn't exist. It would take a series of increasingly ingenious experiments, performed with bacteria and viruses rather than peas and fruit flies, to finally overturn this prejudice and reveal that the genetic material was indeed the seemingly simple molecule that had been hiding in plain sight all along.

The stage was now set for the next act in our drama: the chemical hunt for the molecule of heredity. Scientists knew where to look in the chromosomes but they still had to determine which component of chromosomes actually carried genetic information. The answer, when it finally came, would surprise nearly everyone and would open the door to a revolution in molecular biology that continues to reshape our world today.

Four

The Forgotten Discovery

"Pus, Castles, and a Mysterious Substance"

The discovery of the molecule that would eventually be recognized as the carrier of all genetic information began not with any thought of heredity or genes, but with a young physician's curiosity about the chemical composition of pus. It was 1869, just three years after Mendel had published his unnoticed paper on pea plant inheritance, and the entire science of genetics lay in an unimaginable future. Friedrich Miescher, a twenty-five-year-old Swiss doctor with poor hearing that had discouraged him from clinical practice, had turned instead to laboratory research, seeking to understand life through chemistry rather than through patient care.

Miescher had secured a position in the laboratory of Felix Hoppe-Seyler at the University of Tübingen in Germany, working in a converted castle that had been transformed into one of Europe's leading centers for biochemical research. Hoppe-Seyler was a pioneer in the emerging field of biochemistry, convinced that understanding life required understanding the chemical substances that composed living tissue. He encouraged his students to isolate and characterize the chemical components of cells, and he gave Miescher an appropriately humble starting material: pus-soaked bandages collected from a nearby surgical clinic.

The choice of pus as research material wasn't as strange as it might sound to modern ears. Pus was nothing more than a collection of white blood cells that had died fighting infection, and white blood cells were conveniently available in large quantities from the surgical wards where infection was a constant companion. More importantly, white blood cells had very large nuclei relative to their overall size, making them ideal for anyone interested in studying nuclear contents. Miescher's goal was to isolate and analyze the proteins that composed these nuclei, adding another piece to the growing chemical catalog of cellular components.

An Unexpected Substance

Working in his makeshift laboratory in the old castle's cellar, Miescher developed a careful procedure for separating white blood cells from the bandages, then breaking open these cells to isolate their nuclei. The work required patience and precision he had to keep everything cold to prevent the degradation of cellular components, and he had to work quickly once cells were broken open. When he finally succeeded in isolating relatively pure nuclei and began analyzing their chemical composition, he found what he expected: proteins, present in abundance.

But he also found something else, something that didn't fit into any known category of biological molecules. This mysterious substance had properties unlike any protein Miescher had ever encountered. It was unusually acidic, a characteristic revealed by its behavior in various chemical tests. Most strikingly, it contained a remarkably high proportion of phosphorus, an element that was rare or absent in most biological molecules known at the time. When Miescher tried to dissolve this substance in the weak acid solutions that would normally dissolve proteins, it remained stubbornly insoluble. Yet when he added alkali—a base that would make the solution less acidic the substance readily dissolved.

Miescher realized he had discovered something new, a previously unknown component of cell nuclei. With the straightforward naming convention typical of nineteenth-century scientists, he called it "nuclein" the substance from the nucleus. He knew immediately that it was significant, though he couldn't have imagined just how significant. "It seems probable to me," he wrote in his notes, "that a whole family of such slightly varying phosphorous-rich substances will appear, which will be different from proteins and deserve their own designation as a group."

His supervisor Hoppe-Seyler was skeptical. A mysterious new class of substances? Highly acidic material in cell nuclei? It seemed improbable, perhaps even a result of contamination or experimental error. Hoppe-Seyler, in a display of scientific caution that delayed publication but ultimately strengthened the discovery's credibility, insisted on repeating Miescher's experiments himself before allowing publication. For two years, Hoppe-Seyler worked to verify the findings, eventually confirming that Miescher had indeed discovered something genuinely new.

When the paper finally appeared in 1871, it attracted modest attention in the small world of biochemists but made no great splash. Nuclein was interesting, certainly, but what role did it play in the cell? What function did it serve? These questions remained unanswered, and in the absence of clear functional significance, nuclein remained a biochemical curiosity rather than a molecule of obvious importance.

The Chemical Puzzle

Miescher himself continued to work with nuclein for the rest of his career, which was tragically cut short when he died of tuberculosis at age fifty-one. He shifted to using salmon sperm as his source material, a choice that would prove prescient. Salmon sperm cells are almost nothing but nucleus they're basically DNA delivery vehicles, carrying genetic information and little else making them an extraordinarily rich source of nuclein. Working with kilogram quantities of salmon sperm heads obtained from fisheries on the Rhine river, Miescher was able to isolate nuclein in amounts large enough for detailed chemical analysis.

His analyses revealed that nuclein was unlike anything else in the cell. It wasn't a protein, though it often appeared associated with proteins in the nucleus. It wasn't a lipid or a carbohydrate. It was something fundamentally different, a new category of biological molecule. Miescher could determine its elemental composition it contained carbon, hydrogen, oxygen, nitrogen, and that distinctive phosphorus but the way these elements were arranged and bonded together remained a mystery that the analytical techniques of the 1870s couldn't solve.

Other chemists began working with nuclein, refining isolation procedures and attempting to understand its structure. By the 1880s, researchers had determined that nuclein from different sources had somewhat different properties, leading to the recognition that there were actually two types of nuclein. One type, found abundantly in cell nuclei, would eventually be called deoxyribonucleic acid (DNA). The other type, present in both nuclei and cytoplasm, would be called ribonucleic acid (RNA). The difference between them a single oxygen atom present in RNA but absent in DNA seemed minor at the time but would prove to be profoundly important.

Breaking Down the Molecule

The real breakthrough in understanding nuclein's chemical structure came in the early twentieth century through the painstaking work of Phoebus Levene, a Russian-born biochemist working at the Rockefeller Institute in New York. Levene was a master of chemical analysis, capable of taking complex molecules apart piece by piece and identifying each component with meticulous precision. Over decades of work beginning in the 1900s, he systematically degraded nucleic acids using various chemical and enzymatic treatments, then isolated and identified the fragments.

What emerged from this molecular dissection was a picture of nucleic acids as polymers long chains built from smaller repeating units. Levene identified three types of components that made up these units. First, there was a five-carbon sugar molecule: ribose in RNA, and a slightly modified version called deoxyribose (meaning ribose minus one oxygen atom) in DNA. Second, there were nitrogen-containing compounds called bases, which came in two structural varieties. The larger bases, called purines, were adenine and guanine. The smaller bases, called pyrimidines, were cytosine and thymine in DNA, or cytosine and uracil in RNA. Third, there were phosphate groups chemical units containing phosphorus bonded to oxygen atoms that linked the sugars together into a long chain.

Levene proposed that these components assembled into fundamental units he called nucleotides. Each nucleotide consisted of one sugar molecule bonded to one phosphate group and one base. The phosphate groups formed the backbone of the nucleic acid chain, linking one sugar to the next in a repeating pattern, while the bases stuck out like pendants on a necklace. It was elegant chemistry, carefully worked out through years of rigorous experimentation.

But Levene made a critical error in interpretation that would haunt the field for decades. Based on his analyses, he concluded that nucleic acids contained equal amounts of all four bases. Every sample of DNA, he believed, had exactly as much adenine as guanine, as much cytosine as thymine. This led him to propose what became known as the "tetra nucleotide hypothesis" the idea that DNA was simply a monotonous repetition of a four-nucleotide unit containing one of each base. If this were true, DNA would be like a simple repeating pattern in wallpaper, the same motif over and over again, lacking any meaningful variation or complexity.

This hypothesis had devastating consequences for how scientists thought about DNA. If DNA was just a boring repetition of the same four-unit pattern, it couldn't possibly encode the vast diversity of genetic information. It would be like trying to write all the books in a library using only the same four-word sentence repeated endlessly. The conclusion seemed inescapable: DNA must serve some structural or housekeeping role in chromosomes, but the real genetic material, the substance that carried the complex information needed to specify the characteristics of organisms, must be proteins.

After all, proteins had everything DNA seemed to lack. They were built from twenty different amino acids rather than just four bases. They could fold into an enormous variety of three-dimensional shapes, each with unique properties and functions. They were diverse, complex, and sophisticated—exactly what genetic material ought to be. The prejudice in favor of proteins became so strong that even when evidence began to accumulate suggesting DNA's importance, many scientists dismissed or rationalized it away, clinging to their conviction that genes must be made of protein.

The Stubborn Molecule

Despite this institutional bias, DNA refused to remain in the background. Chemical analyses in the 1920s and 1930s revealed that DNA was not uniformly distributed throughout the cell but was concentrated specifically in the nucleus, precisely where chromosomes resided and where, according to the chromosomal theory of inheritance, genes should be located. Moreover, the amount of DNA in cells followed a suggestive pattern: body cells that contained two sets of chromosomes had twice as much DNA as sperm or egg cells that contained only one set. This correlation between chromosome number and DNA content hinted at a deeper connection.

Some researchers noticed other intriguing facts about DNA. It was remarkably stable, persisting unchanged through the life of a cell and being faithfully duplicated when cells divided. It had a constant amount in all body cells of a given species every human liver cell, skin cell, or neuron contained the same quantity of DNA, despite these cells having vastly different structures and functions. Even more tellingly, different species had different amounts of DNA, with more complex organisms generally having more DNA than simpler ones. These observations were suggestive but not definitive. They could be interpreted as evidence that DNA was the genetic material, or they could be explained away as merely reflecting DNA's role as a structural scaffold for the real genetic material, the proteins.

The scientific community's bias toward proteins was so powerful that it took increasingly direct and irrefutable evidence to begin changing minds. That evidence would come not from studying normal heredity in plants or animals, but from investigating a bizarre and seemingly magical phenomenon discovered in bacteria: the ability of dead bacteria to permanently transform living bacteria, conferring new hereditary characteristics through some mysterious "transforming principle" that could pass from one cell to another carrying genetic instructions.

This discovery, made by a British physician studying pneumonia-causing bacteria in 1928, would set in motion a chain of investigations that would ultimately force scientists to confront the evidence they had been avoiding. The transforming principle experiments would demonstrate, through a process of systematic elimination, that DNA not protein was the substance that carried genetic information. But proving this would require sixteen years of painstaking work by a team of scientists at the Rockefeller Institute, fighting against both technical challenges and the entrenched skepticism of their colleagues.

The story of how DNA emerged from obscurity to claim its rightful place as the molecule of heredity is one of the great detective stories in science a tale of careful experiments, creative thinking, and the gradual triumph of evidence over assumption. It would take the work of many scientists over several decades, building piece by piece, experiment by experiment, until the conclusion became unavoidable. The genetic material was not the complex, sophisticated proteins that everyone expected. It was the supposedly simple, monotonous nucleic acid that had been sitting in plain sight all along, waiting for someone to recognize its true significance.

Five

The Transforming Principle

"Dead Bacteria Tell Tales"

The year was 1928, and pneumonia was a killer. Before the development of antibiotics, bacterial pneumonia cut down the young and old, the weak and strong, with brutal efficiency. Physicians could do little more than keep patients comfortable and hope that their immune systems would mount a successful defense. In this context, understanding the bacteria that caused pneumonia Streptococcus pneumonia was not merely an academic exercise but a matter of life and death.

Frederick Griffith was a medical officer working at the Ministry of Health in London, a quiet, methodical man more comfortable with microscopes and bacterial cultures than with human patients. He had spent years studying pneumococcal bacteria, trying to understand why some strains caused deadly disease while others were harmless. His work required patience and meticulous attention to detail, qualities that Griffith possessed in abundance. He was not looking for a revolutionary discovery about the nature of genetic material he was simply trying to understand bacterial virulence. But in science, as in life, the most important discoveries often come when we're looking for something else entirely.

The Two Strains

Griffith worked with two distinct strains of Streptococcus pneumonia that could be easily distinguished under the microscope and by their effects on laboratory mice. The first strain, designated "S" for smooth, produced bacterial colonies with a glossy, smooth appearance when grown on laboratory plates. This smooth appearance came from a polyacrylamide capsule that surrounded each bacterial cell like a protective coating. The capsule was more than just cosmetic it was essential for the bacteria's virulence, its ability to cause disease. The smooth, capsule-coated bacteria could evade the mouse immune system, multiply unchecked in the bloodstream, and kill a mouse within days of infection.

The second strain, designated "R" for rough, produced colonies with a rough, irregular appearance. These bacteria had lost the ability to make the polyacrylamide capsule through a mutation, and without this protective coating, they were essentially harmless. A mouse could be injected with millions of R bacteria and suffer no ill effects—the immune system would easily recognize and destroy the unprotected invaders. The difference between life and death, between virulence and harmlessness, came down to the presence or absence of that polyacrylamide capsule.

This much was straightforward biology, and Griffith confirmed it through systematic experiments. Inject mice with living S bacteria the mice die and their blood teems with S bacteria. Inject mice with living R bacteria the mice survive and remain healthy. Heat-kill the S bacteria before injection the mice survive, because dead bacteria can't cause disease. These results were entirely expected and raised no particular questions.

But Griffith performed one more experiment, almost as a control, that would shake the foundations of biology. He took heat-killed S bacteria completely dead, incapable of growing or reproducing and mixed them with living R bacteria. Neither component alone could harm a mouse: the S bacteria were dead, and the R bacteria were harmless. Logically, the mixture should be harmless as well. When Griffith injected this mixture into mice, he expected the mice to survive.

Instead, the mice died.

The Impossible Result

Griffith's first thought must have been that he had made an error, that some of the S bacteria hadn't actually been killed by the heat treatment. This was the sort of technical mistake that could easily happen in a busy laboratory. But when he examined the blood of the dead mice under his microscope, he saw something that couldn't be explained by incomplete heat treatment. The blood was full of living S bacteria smooth, capsule-coated, virulent bacteria. Not just a few survivors that had escaped the heat treatment, but massive numbers of bacteria, all with the smooth phenotype.

More remarkable still, when Griffith isolated these S bacteria from the dead mice and grew them in culture, they continued to produce S-type offspring. This wasn't a temporary change or a reversion to an ancestral type it was a permanent, heritable transformation. Somehow, the living R bacteria had acquired the ability to make capsules, transforming themselves into S bacteria. This transformation passed to their offspring and remained stable through countless generations. The R bacteria had permanently gained a new genetic capability.

The result was reproducible. Griffith performed the experiment again and again, always with the same outcome. Living R bacteria plus heat-killed S bacteria equals dead mice and living S bacteria. He tried various controls, checking whether the transformation might occur without the R bacteria or without the dead S bacteria. It didn't. Both components were necessary. The heat-killed S bacteria contained something some "transforming principle," as Griffith called it that could enter living R bacteria and permanently change their hereditary characteristics, conferring upon them the ability to synthesize the polyacrylamide capsule.

Griffith published his findings in 1928 in the Journal of Hygiene. The paper was careful and thorough, presenting the data without wild speculation about its implications. But the implications were nonetheless revolutionary. Something in the dead S bacteria something that survived heat treatment that killed the bacteria themselves could carry genetic information. This something could enter living bacteria and become incorporated into their hereditary apparatus, permanently altering the genes they passed to their offspring. Griffith had discovered what we would now call horizontal gene transfer, the movement of genetic information between organisms outside of normal reproduction.

The Skeptical Reception

The scientific community's reaction to Griffith's discovery was a mixture of fascination and skepticism. Some researchers found the results intriguing and began their own transformation experiments, confirming that the phenomenon was real. Others suspected artifacts, contamination, or misinterpretation. The idea that bacteria could acquire new hereditary characteristics from dead bacteria seemed to verge on the mystical, reminiscent of discredited theories about the inheritance of acquired characteristics.

Even those who accepted the reality of transformation puzzled over its mechanism. What was this transforming principle? Was it truly genetic material, actually changing the bacteria's genes, or was it something else—perhaps an enzyme that activated a per-existing but dormant genetic capability? The latter interpretation seemed more palatable to many scientists, as it didn't require accepting the radical notion that genetic material could be extracted from dead cells and transferred to living ones.

Griffith himself continued his research but made little progress in identifying the chemical nature of the transforming principle. He was a bacteriologist, not a biochemist, and he lacked the training and equipment needed to isolate and purify the active substance from dead bacteria. The question of what the transforming principle actually was would require someone with different expertise, someone who combined biochemical sophistication with the patience to pursue a project that might take years or decades to complete.

That someone was working at the Rockefeller Institute in New York City. Oswald Avery was already in his fifties when he learned of Griffith's transformation experiments, but age had not dimmed his scientific curiosity or his determination. Avery recognized immediately that identifying the transforming principle could answer one of biology's most fundamental questions: what are genes made of?

Avery's Quest Begins

Oswald Theodore Avery was, in many ways, an unlikely revolutionary. Born in 1877 in Nova Scotia, Canada, he had trained as a physician but quickly turned to laboratory research, finding clinical practice distasteful. He joined the Rockefeller Institute in 1913, where he devoted himself to studying pneumococcal bacteria with the same meticulous care and attention to detail that had characterized Griffith's work. Quiet and unassuming, Avery was the sort of scientist who preferred hard data to speculation, careful experimentation to dramatic pronouncements.

When Avery first read Griffith's 1928 paper, he was skeptical. Bacterial transformation seemed too strange, too improbable. But subsequent experiments by other researchers confirmed the phenomenon, and Avery's skepticism transformed into intense curiosity. If transformation was real and it appeared to be then understanding it could reveal fundamental truths about heredity. Avery decided to devote his laboratory's resources to identifying the transforming principle, whatever it might be.

The task Avery set himself was daunting. He needed to extract the transforming principle from dead S bacteria in pure form, determine its chemical nature, and prove definitively that it was responsible for transformation. This would require developing new purification methods, conducting countless chemical tests, and performing transformation experiments with each purified fraction to determine which component carried the transforming activity. The work would be tedious, time-consuming, and fraught with technical challenges.

Avery began the work in the late 1920s with his colleague Colin McLeod, and they were later joined by Maclyn McCarty, a young physician-scientist. Together, they embarked on what would become a sixteen-year quest to answer a single question: what is the transforming principle made of?

The Process of Elimination

The approach Avery and his colleagues took was systematic and rigorous. They would purify the transforming principle through increasingly refined extraction and purification steps, testing the transforming activity at each stage. Then they would subject their preparations to various chemical and enzymatic treatments designed to destroy specific types of molecules proteins, lipids, RNA, or DNA and test whether transformation still occurred. If destroying a particular type of molecule eliminated transformation, that molecule must be the transforming principle.

The first experiments focused on proteins, since most scientists assumed that genes must be made of protein. Avery's team treated their preparations with proteolytic enzymes enzymes that specifically digest and destroy proteins. They used trypsin, chymotrypsin, and other protein-degrading enzymes, subjecting their extracts to conditions that would utterly destroy any protein present. Then they tested the treated material for its ability to transform R bacteria into S bacteria.

Transformation continued unabated. Whatever the transforming principle was, it wasn't protein, or at least protein wasn't the essential component. This result was surprising but not yet definitive. Perhaps their preparations still contained some protein that had escaped the enzyme treatment. Avery needed to push further, to purify the transforming principle more completely and test other possibilities.

They treated their preparations with lipid-dissolving solvents to remove any fatty substances. Transformation continued. They used enzymes that specifically destroyed RNA. Transformation continued. With each negative result each treatment that failed to eliminate transformation the possibilities narrowed. The transforming principle was remarkably resistant to degradation, surviving treatments that destroyed most biological molecules.

Finally, they obtained highly purified preparations of the transforming principle and subjected them to detailed chemical analysis. The results were unambiguous. The preparations contained DNA lots of it and virtually no protein. The ratio of nitrogen to phosphorus matched what would be expected for pure DNA. Tests for protein came back negative or showed only trace amounts that could be dismissed as contamination. The transforming principle looked like DNA, behaved like DNA, and had the chemical composition of DNA.

But chemical similarity wasn't proof. Avery needed to demonstrate that destroying DNA would destroy the transforming activity. The problem was that in the early 1940s, when this work was being completed, enzymes that specifically destroyed DNA were rare and poorly characterized. Avery and McCarty searched for such an enzyme and eventually found one that showed specific DNase activity the ability to break down DNA while leaving other molecules intact.

When they treated their transforming principle preparations with this DNA-destroying enzyme, transformation stopped completely. The material that had reliably transformed R bacteria into S bacteria through hundreds of previous experiments became inert, incapable of causing any transformation whatsoever. The conclusion seemed inescapable: DNA was the transforming principle. DNA carried genetic information. DNA, not protein, was the genetic material.

The 1944 Announcement

In February 1944, Avery, MacLeod, and McCarty submitted their paper to the Journal of Experimental Medicine. The paper was a masterpiece of scientific writing careful, thorough, and measured in its conclusions. They presented sixteen years' worth of data demonstrating that the transforming principle was DNA. They showed their purification methods, their chemical analyses, their enzymatic degradation experiments, all pointing to the same conclusion. Yet they remained cautious in their interpretations, acknowledging that absolute proof was difficult and that their conclusions might need revision in light of future evidence.

"The evidence presented supports the belief that a nucleic acid of the desoxyribose type is the fundamental unit of the transforming principle of Pneumococcus Type III," they wrote, using the older spelling "desoxyribose" for what we now call deoxyribose. It was careful language, hedged with appropriate scientific caution, but the message was clear: DNA is the genetic material.

The paper should have caused an immediate sensation. It should have overturned the protein-gene hypothesis and redirected the entire field of genetics toward DNA. But it didn't, at least not immediately. The scientific community's reaction was strangely muted, a mixture of interest and persistent skepticism. Some scientists accepted Avery's conclusion and began reorienting their research accordingly. Many others remained doubtful, finding reasons to question or dismiss the findings.

The skepticism took various forms. Some argued that the DNA preparations might still contain tiny amounts of protein that were doing the actual transformation, with DNA serving merely as a carrier. Others suggested that DNA might be activating per-existing but dormant genes rather than actually transferring new genetic information. Still others simply couldn't abandon their conviction that genes must be made of protein and treated Avery's work as interesting but probably wrong or at least incomplete.

Part of the resistance came from the persistence of Levine's tetra nucleotide hypothesis. If DNA was just a boring repetition of the same four-nucleotide sequence, how could it carry genetic information? How could it specify different genes for different traits? The answer that Levine was wrong about DNA's structure, that DNA sequences were not monotonous repetitions but varied in complex and meaningful ways had not yet been established. Until someone demonstrated that DNA had sequence variation and complexity, the protein hypothesis remained viable in many scientists' minds.

Avery himself was disappointed but not surprised by the tepid reception. He knew he had done rigorous work, and he was confident in his conclusions, but he also understood that overturning established scientific paradigms required overwhelming evidence and often took time. He continued to work on bacterial transformation until his retirement, but he never pursued the Nobel Prize-level recognition that his discovery deserved. When he died in

1955, he had still not received the prize, though many believed he deserved it.

The scientific community needed more evidence, something so clear and unambiguous that even the most stubborn skeptics would have to concede that DNA was the genetic material. That evidence would come eight years after Avery's paper, from an experiment so elegantly simple in its design and so unequivocal in its results that it would finally lay the protein hypothesis to rest. All it required was some bacteriophages, a bit of radioactive isotopes, and a common kitchen appliance a Waring blender.

Six

The Blender Experiment

"Viruses, Radioactivity, and a Kitchen Appliance"

Alfred Hershey was not a man given to scientific theatrics or dramatic pronouncements. Colleagues described him as quiet, almost reclusive, more comfortable with experiments than with people. He worked at the Carnegie Institution's Department of Genetics at Cold Spring Harbor Laboratory on Long Island, New York, in a community of scientists who shared his passion for understanding the mechanisms of heredity at the molecular level. Hershey had spent years studying bacteriophages viruses that infect bacteria and by 1951 he had become deeply interested in a question that seemed, on the surface, almost absurdly simple: when a phage infects a bacterium, what actually enters the cell?

The question wasn't trivial. Understanding what entered the bacterial cell during infection would reveal what constituted the viral genetic material. Bacteriophages, or phages as they were commonly called, were ideal organisms for asking this question. They had an uncomplicated structure that had been revealed through electron microscopy and biochemical analysis.

A phage particle consisted essentially of two components: an outer protein coat that formed an elaborate structure resembling a lunar landing module, complete with a head, tail, and spidery tail fibers, and an inner core of DNA packed tightly inside the head. The phage's life cycle was simple and brutal. The virus would attach to a bacterial cell, inject something into the cell, and approximately twenty minutes later, the bacterium would burst open and release hundreds of new phage particles, each identical to the original infecting virus.

The question was straightforward: did the protein coat enter the cell, did the DNA enter the cell, or did both enter? The answer would definitively identify the genetic material. Whatever went into the cell and directed the synthesis of new phages must be the carrier of genetic information. If protein entered, genes were made of protein. If DNA entered, genes were made of DNA. If both entered, the situation would remain ambiguous. Hershey needed an experiment that could distinguish between these possibilities with absolute clarity.

A Brilliant Partnership

Martha Chase entered Hershey's laboratory in 1950 as a research assistant, fresh from graduate school and eager to contribute to the exciting work happening at Cold Spring Harbor. The laboratory was a hub of phage research, part of what was known as the "Phage Group" an informal network of scientists pioneered by Max Delbrück who had chosen bacterial viruses as their model system for understanding fundamental questions in genetics. Chase brought energy, technical skill, and fresh perspectives to Hershey's meticulous research program.

The collaboration between Hershey and Chase proved remarkably productive. They made an effective team: Hershey with his deep knowledge of phage biology and his preference for experiments that gave unambiguous answers, and Chase with her experimental dexterity and willingness to tackle technically challenging procedures. Together they designed what would become one of the most famous experiments in molecular biology, an experiment so clear in its logic and so definitive in its results that it would be taught to biology students for generations to come.

The experimental design exploited a fundamental difference in the chemical composition of proteins and DNA. Proteins are built from amino acids, some of which contain sulfur atoms but none of which contain phosphorus. DNA, by contrast, contains abundant phosphorus in the sugar-phosphate backbone that links nucleotides together, but it contains no sulfur. This difference provided a perfect opportunity for labeling. If phages could be grown in the presence of radioactive sulfur-35, the radioactive sulfur would be incorporated specifically into proteins. If phages were grown with radioactive phosphorus-32, the radioactive phosphorus would be incorporated specifically into DNA. Using radioactivity as a tag, Hershey and Chase could track where proteins went and where DNA went during infection.

The first challenge was growing phages that contained the radioactive labels. This required infecting bacteria that had been grown in media containing either radioactive sulfur or radioactive phosphorus, then allowing the phages to reproduce and incorporate the radioactive isotopes into their structure. Hershey and Chase performed these infections carefully, producing two batches of phages: one with radioactively labeled proteins (^{35}S-phages) and one with radioactively labeled DNA (^{32}P-phages). The radioactivity made these phages behave no differently from normal phages they infected bacteria with normal efficiency and produced normal offspring but now their components could be tracked with exquisite sensitivity using instruments that detected radioactive decay.

The Experiment Unfolds

The actual infection experiment was elegantly straightforward in concept, though technically demanding in execution. Hershey and Chase allowed their radioactively labeled phages to attach to bacterial cells and begin infection. The phages would stick to the bacterial surface, inject their genetic material, and begin the process that would eventually produce hundreds of new phages. But Hershey and Chase didn't wait for the infection to complete. Instead, just a few minutes after infection began long enough for injection to occur but before the bacteria lased they subjected the mixture to violent agitation in a Waring blender.

The blender treatment had a specific purpose: to shear off the empty phage coats that remained attached to the bacterial surface after injection. Imagine the phages as tiny syringes injecting their contents into bacteria. After injection, the syringe remains stuck to the bacterial surface, empty but still attached. The blender's vigorous shaking ripped these empty "syringes" off the bacterial surface without killing the bacteria or disrupting the infection process that was already underway inside the cells.

After blending, Hershey and Chase used centrifugal to separate the bacterial cells from the detached phage coats. Bacteria, being larger and heavier than the phage debris, would pellet at the bottom of the centrifuge tube when spun at high speed, while the empty phage coats remained in the liquid above. This separation was crucial: it allowed Hershey and Chase to ask whether the radioactive label stayed with the bacteria or remained with the phage coats.

The results were unambiguous and deeply satisfying in their clarity. When they performed the experiment with ^{35}S-labeled phages phages with radioactive sulfur in their proteins the radioactivity remained almost entirely in the liquid fraction containing the detached phage coats. Very little radioactivity was found associated with the bacterial pellet. The protein coats had stayed outside the cell, sheared off by the blender and separated by centrifugal. This made sense: the protein formed the external structure of the phage, and it remained external to the bacterium during infection.

When they repeated the experiment with ^{32}P-labeled phages phages with radioactive phosphorus in their DNA the results were strikingly different. Now the radioactivity was found primarily in the bacterial pellet. The DNA had entered the bacteria during infection and remained inside even after the blender treatment separated the bacteria from the empty phage coats. Furthermore, when they allowed these infected bacteria to complete their life cycle and burst open, releasing new phages, the radioactive phosphorus appeared in the progeny phages. The DNA that entered the bacteria had been used to make new phages, demonstrating that it carried the information needed to direct viral reproduction.

The Implications Become Clear

The logic was inescapable. The material that entered the bacterial cell and directed the production of new phages was DNA, not protein. The protein coat remained outside, serving merely as a protective shell and an attachment mechanism. All the genetic information needed to make new phages instructions for synthesizing phage proteins, assembling phage structures, and reproducing the phage DNA itself resided in the DNA that was injected into the cell.

Hershey and Chase published their results in 1952 in the Journal of General Physiology under the straightforward title "Independent Functions of Viral Protein and Nucleic Acid in Growth of Bacteriophage." The paper was characteristically understated, presenting the data clearly but without dramatic claims or philosophical speculation. The key finding was stated matter-of-factly: "When bacteria are infected with bacteriophage containing radioactive DNA, and the infected bacteria are then agitated violently to remove the phage coats, 75% to 80% of the DNA remains with the infected bacteria and is transmitted to the phage progeny."

The scientific community's reaction this time was dramatically different from the response to Avery's work eight years earlier. The Hershey-Chase experiment was so direct, so unambiguous, so free from possible alternative interpretations that skepticism crumbled. Here was visual, quantitative proof that DNA entered cells while protein did not, that DNA was transmitted to progeny while protein was not, that DNA carried genetic information while protein served structural and protective roles.

The experiment's beauty lay partly in its simplicity. Anyone could understand the basic logic: label the two components differently, see which one enters the cell, and that component must be the genetic material. The blender was a particularly inspired touch a common kitchen appliance used to answer one of biology's most fundamental questions. Science journalists loved this detail, and the "Waring Blender Experiment" became famous both within and beyond the scientific community.

The timing proved crucial as well. By 1952, the climate in molecular biology had shifted. Avery's work had planted seeds of doubt about the protein hypothesis, even among those who remained officially skeptical. Other experiments had accumulated showing DNA's importance, including work by André Boivin in France demonstrating that DNA content correlated precisely with chromosome number across species. The Hershey-Chase experiment arrived at a moment when the scientific community was ready to abandon the protein hypothesis if given sufficient reason. Hershey and Chase provided that reason.

The Resistance Crumbles

Some diehards still clung to alternative explanations. Perhaps a tiny amount of protein entered along with the DNA and wasn't detected by the radioactive labeling. Perhaps DNA catalyzed the synthesis of proteins using information stored elsewhere. Perhaps the experiment worked for phages but wouldn't hold true for other organisms. But these objections grew increasingly desperate and unconvincing. The evidence was mounting from multiple directions, all pointing toward the same conclusion: DNA was the genetic material, the molecule that carried hereditary information from one generation to the next.

The combination of Avery's transformation experiments and the Hershey-Chase blender experiment convinced the vast majority of molecular biologists. By 1952, the question of what genes were made of had been answered. The protein hypothesis, which had dominated thinking for decades based on assumptions about complexity and sophistication rather than experimental evidence, was dead. DNA, the supposedly simple and monotonous molecule that biochemists had dismissed as too boring to be important, stood revealed as the carrier of life's instructions.

But answering one question immediately raised a host of others. If DNA was the genetic material, how did it store information? The structure of DNA must somehow encode genetic instructions, but how? DNA was a long polymer built from just four types of nucleotide how could such a molecule specify the thousands of different genes needed to make a functional organism? How did DNA replicate, ensuring that copies of genetic information were accurately transmitted to offspring? How was the information in DNA used to direct the synthesis of proteins, which everyone agreed were the workhorses of cellular function?

These questions demanded knowledge of DNA's three-dimensional structure. The chemical composition of DNA was known Levene had worked that out decades earlier but composition alone couldn't reveal how DNA functioned. Just as understanding how a car works requires knowing not just that it contains metal, rubber, and glass, but how these materials are arranged and connected, understanding how DNA worked would require knowing its precise three-dimensional architecture.

By the early 1950s, several research groups were racing to determine DNA's structure. They knew the structure must be spectacular, must contain within its architecture the key to understanding heredity at the molecular level. They knew that whoever solved the structure would achieve scientific immortality and probably a Nobel Prize. They knew that the answer was tantalizingly close, waiting to be discovered by someone with the right combination of data, insight, and perhaps a bit of luck.

That race to discover DNA's structure would prove to be one of the most dramatic and controversial episodes in the history of science, involving brilliant insights, fierce competition, unauthorized data sharing, and ethical questions that persist to this day. The story would feature some of science's most celebrated figures and some of its most tragic ones, and it would culminate in the revelation of a structure so elegant, so perfect in its simplicity, that it would immediately explain both how genetic information is stored and how it is copied. But before we can understand how that structure was discovered, we need to understand one more piece of the puzzle a pattern hidden in DNA's chemical composition that would prove crucial to unlocking its secrets.

Seven

Chargaff's Rules

"The Hidden Pattern in DNA"

Erwin Chargaff was sitting in his office at Columbia University in New York City in the spring of 1944, reading a copy of the Journal of Experimental Medicine, when he encountered Avery, McLeod, and McCarty's paper describing the transforming principle. Most scientists who read that paper found it interesting but not convincing enough to change their research directions. Chargaff had a different reaction. He was electrified by what he read, immediately grasping the significance of the findings. If DNA truly was the genetic material and Chargaff believed the evidence was compelling then DNA must be far more interesting and complex than anyone had imagined.

This realization changed the course of Chargaff's career. He had been working on various problems in biochemistry, particularly the chemistry of lipids and other biological molecules, but now he decided to devote his laboratory's efforts to understanding DNA. If DNA carried genetic information, its structure and composition must reflect that function. Perhaps hidden within

DNA's seemingly simple four-base composition lay patterns and complexities that had been overlooked. Chargaff determined to look more carefully at DNA's chemical composition than anyone had before, using the most accurate analytical techniques available.

A New Approach to an Old Question

The conventional wisdom, established by Phoebus Levene decades earlier, held that DNA contained equal amounts of all four bases: adenine, guanine, cytosine, and thymine occurred in the same proportions, creating a monotonous, repetitive molecule. This tetra nucleotide hypothesis had become scientific dogma, repeated in textbooks and taught to students, despite being based on relatively crude analytical methods that might not have detected subtle differences in base composition.

4Chargaff suspected that Levene might have been wrong, or at least that his analyses might have been insufficiently precise to detect important variations. The analytical techniques available in the 1940s had improved dramatically since Levene's time, particularly a method called paper chromatography that had been developed during World War II. This technique allowed researchers to separate complex mixtures of similar molecules with unprecedented resolution and accuracy. If DNA from different species or different genes had different base compositions if the proportions of the four bases varied in meaningful ways paper chromatography should be able to detect those differences.

Chargaff assembled a research team and set them to work analyzing DNA from as many different sources as they could obtain. They extracted DNA from bacteria, from yeast, from sea urchins, from salmon, from cattle, and from humans. They treated this DNA with acid to break it down into its constituent bases, then used paper chromatography to separate and quantify each base with meticulous precision. The work was painstaking and time-consuming, requiring careful technique and multiple measurements to ensure accuracy, but Chargaff insisted on the highest standards. If he was going to challenge Levene's tetra nucleotide hypothesis, his data would need to be unassailable.

The Pattern Emerges

The results began coming in during 1949 and 1950, and they revealed something remarkable. Levene had been wrong. DNA did not contain equal amounts of all four bases. Different species had dramatically different base compositions. DNA from tuberculosis bacteria contained roughly 15% adenine, 15% thymine, 35% guanine, and 35% cytosine. DNA from sea urchin sperm had approximately 28% adenine, 28% thymine, 21% guanine, and 21% cytosine. Human DNA fell somewhere in between. The ratios varied widely from species to species, demolishing any notion that DNA was a simple, invariant tetra nucleotide.

This variation was exactly what you would expect if DNA carried genetic information. Different species needed different genetic instructions, so their DNA should have different compositions. The finding supported the idea that DNA was complex enough, variable enough, sophisticated enough to encode genes. But Chargaff noticed something else in his data, something even more intriguing than the species-to-species variation.

Within any given species, despite the variation in overall base composition, certain ratios remained remarkably constant. The amount of adenine always equaled the amount of thymine, within the limits of experimental error. The amount of guanine always equaled the amount of cytosine, again within experimental error. This was true for DNA from every species Chargaff examined, from the simplest bacteria to humans. The rule held universally: A equals T, and G equals C.

Chargaff published his findings in 1950, presenting his data in careful tables that showed the base compositions of DNA from numerous organisms. The paper appeared in the Journal of General Physiology under the title "Chemical Specificity of Nucleic Acids and Mechanism of Their Enzymatic Degradation." It was technical and understated, but the implications were profound. DNA was not a monotonous repetition of the same tetra nucleotide unit it had sequence variation and chemical complexity. Different organisms had different DNA compositions, consistent with DNA encoding different genetic information. And mysteriously, adenine always paired numerically with thymine, guanine with cytosine.

Why should this be? Chargaff himself couldn't fully explain it, though he suspected it must reflect something important about DNA's structure or function. The one-to-one ratios suggested some kind of pairing or complementary between bases, but the nature of that relationship remained unclear. Was it a chemical bonding between complementary bases? A structural requirement of how DNA was organized? Or something else entirely?

What Chargaff had discovered, though he didn't fully realize it at the time, was a crucial clue to DNA's three-dimensional structure. The equal ratios of adenine to thymine and guanine to cytosine weren't coincidences or artifacts they were inevitable consequences of how DNA was built. In the double helix structure that would soon be revealed, adenine on one strand always paired with thymine on the opposite strand, and guanine always paired with cytosine. Chargaff's rules reflected the base pairing that held the two strands of the double helix together. But discovering that structure would require bringing together Chargaff's chemical data with other types of evidence, particularly the patterns revealed by X-ray crystallography.

Chargaff himself became somewhat bitter in later years, feeling that his contribution had been insufficiently recognized. He had provided one of the key pieces of evidence that led to the discovery of DNA's structure, yet when the Nobel Prize was awarded in 1962 for that discovery, his name was not among the recipients. He felt that Watson and Crick, the two men who would propose the double helix model, had not given him proper credit and had been dismissive of his work. Whether this bitterness was justified remains a matter of debate among historians of science, but there's no question that Chargaff's rules were essential to understanding DNA's structure. Without knowing that A equals T and G equals C, the double helix model would have been much harder to deduce.

By 1950, all the pieces needed to solve DNA's structure were scattered across different laboratories, different continents, and different scientific disciplines. The chemical composition was known. The transforming principle experiments had proven that DNA carried genetic information. Chargaff's rules had revealed the mysterious one-to-one ratios of bases. X-ray crystallography studies, still being refined, were producing diffraction patterns that contained encoded information about DNA's three-dimensional architecture. What was needed now was someone to gather all these pieces together, to see how they fit, and to construct a model of DNA's structure that would explain all the experimental observations while also revealing how DNA could store, replicate, and transmit genetic information.

That synthesis was about to happen, but it would occur in an atmosphere of intense competition, questionable ethics, and scientific drama that would generate controversy extending far beyond the discovery itself. The race to discover DNA's structure would involve four main competitors working in three different laboratories, and it would ultimately produce one of the most famous scientific papers ever written a 900-word letter to Nature that changed biology forever.

Eight

The Players Assemble

"Four Groups, One Prize"

In the early 1950s, the international scientific community understood that determining DNA's three-dimensional structure was the next great prize in molecular biology. Whoever solved the structure would instantly achieve scientific immortality, would almost certainly win a Nobel Prize, and would be remembered as having made one of the fundamental discoveries about life itself. The stakes could hardly have been higher, and several research groups around the world were racing toward the answer.

The competition centered on three main locations, each with its own advantages, personalities, and approaches. At the California Institute of Technology in Pasadena, the legendary Linus Pauling brought his unmatched understanding of chemical bonding and structural chemistry to bear on the problem. At King's College London, Maurice Wilkins and Rosalind Franklin used X-ray crystallography to probe DNA's structure with unprecedented precision, though their collaboration was marked by tension and miscommunication. And at Cambridge University, two young scientists James Watson, a brash American postdoctoral researcher, and Francis Crick, a graduate student who talked too much and had yet to complete his Ph.D. were taking a different approach entirely, building physical models and trying to synthesize insights from everyone else's work into a coherent structural picture.

Linus Pauling: The Favorite

If you had asked scientists in 1951 who would solve DNA's structure, most would have said Linus Pauling without hesitation. Pauling was already a giant of twentieth-century chemistry, having made fundamental contributions to understanding chemical bonding and molecular structure. He had recently scored a spectacular success by determining the alpha helix structure of proteins, using a combination of model building, chemical principles, and insight to deduce a structure that turned out to be precisely correct. The alpha helix discovery had made headlines and reinforced Pauling's reputation as the world's preeminent structural chemist. DNA seemed like a natural next target for his talents.

Pauling had several advantages. He understood chemical bonding at a deep, intuitive level that few other scientists could match. He was brilliant at model building, constructing physical representations of molecules using balls and sticks, then manipulating these models to see what arrangements were chemically reasonable. He had excellent graduate students and postdocs working in his laboratory, providing technical support and fresh perspectives. And he had the confidence some might say arrogance that comes from a long string of successful predictions and discoveries.

But Pauling also faced obstacles, some of which he didn't fully appreciate. He was working at Cal tech in California, far from the X-ray crystallography labs in England where the best experimental data on DNA's structure was being collected. More seriously, Pauling's political activities and left-leaning sympathies had made him a target during the McCarthy era of communist witch hunts. In 1952, the U.S. State Department revoked his passport, preventing him from traveling abroad to conferences where he might have learned about the latest experimental findings on DNA. This isolation from the mainstream of DNA research would prove costly.

King's College London: The Experimentalists

At King's College in London, the physical chemistry laboratory possessed what was arguably the world's most sophisticated X-ray crystallography operation, specifically focused on biological molecules. X-ray crystallography was a powerful technique for determining molecular structure, based on the principle that X-rays passing through a crystallized substance would be diffracted bent and scattered in patterns that reflected the three-dimensional arrangement of atoms in the crystal. By analyzing these diffraction patterns mathematically, researchers could work backwards to deduce the structure that must have produced them.

Maurice Wilkins had been working with DNA at King's College since the late 1940s, using X-ray crystallography to probe its structure. Wilkins was a careful, methodical scientist, a physicist by training who had transitioned to biology after working on the Manhattan Project during World War II. He found DNA fascinating and believed that determining its structure would be tremendously important, but he approached the problem patiently, without the sense of desperate urgency that characterized some of his competitors.

In January 1951, Rosalind Franklin joined Wilkins at King's College, recruited to bring her expertise in X-ray crystallography to bear on the DNA problem. Franklin was a brilliant physical chemist who had spent three productive years in Paris perfecting her X-ray crystallography techniques. She was meticulous, demanding the highest standards from herself and others, and intensely focused on extracting every possible piece of information from her experimental data. She was also independent-minded and uncompromising, unwilling to speculate beyond what her data could support.

From the beginning, the relationship between Wilkins and Franklin was troubled. There had been miscommunication about Franklin's role Wilkins thought she was coming to assist his DNA research, while Franklin believed she had been given DNA as her own independent project. This fundamental misunderstanding poisoned their professional relationship. They worked in the same building, on the same problem, but largely independently and with minimal communication. Franklin focused on obtaining the best possible X-ray diffraction photographs of DNA and interpreting them rigorously, while Wilkins pursued his own investigations and increasingly looked to collaborators elsewhere, particularly at Cambridge.

Cambridge: The Model Builders

At the Cavendish Laboratory in Cambridge, a very different approach to the DNA problem was being pursued. James Watson and Francis Crick were not experimentalists in the traditional sense they weren't generating new data about DNA through laboratory experiments. Instead, they were synthesizers, trying to assemble everyone else's findings into a coherent three-dimensional structure. Their method was model building: constructing physical models from metal plates and rods, testing whether proposed structures were consistent with known chemical principles and experimental data, and using the models to generate new insights about how DNA might be organized.

James Watson had arrived in Cambridge in 1951 as a young American postdoctoral researcher, only twenty-three years old and already convinced that DNA's structure was the most important problem in biology. He had been trained in phage genetics and had attended a conference in Naples where he had seen Wilkins present an X-ray diffraction photograph of DNA. That image had electrified Watson, showing him that DNA had a regular, crystalline structure that could potentially be solved. He had maneuvered his way to Cambridge specifically to work on DNA structure, though his official fellowship was supposed to support research on something else entirely.

Francis Crick was twelve years older than Watson, but he was still a graduate student, having started his Ph.D. relatively late after working on magnetic mines during World War II. Crick was brilliant, gregarious, and notorious for his booming laugh and tendency to dominate conversations with enthusiastic theorizing about whatever scientific problem caught his interest. He had been working on protein structure but became increasingly fascinated by DNA after Watson arrived. The two men formed an instant partnership, complementary in their strengths and united in their conviction that DNA's structure could be solved through model building combined with careful attention to everyone else's experimental findings.

Their approach was audacious bordering on arrogant. They weren't doing the painstaking experimental work themselves they were essentially trying to solve the structure by thinking about it cleverly and building models based on other people's data. This didn't endear them to the experimentalists, particularly at King's College, who felt that Watson and Crick were trying to scoop their work without doing the hard laboratory research. The relationship between Cambridge and King's College became increasingly strained as Watson and Crick pushed forward with their model building while Franklin and Wilkins continued their careful experimental studies.

The Rules of Engagement

The scientific competition was complicated by informal agreements and British politeness about not stepping on others' research territories. After an early, disastrous model-building attempt by Watson and Crick in 1951 they built a structure that Franklin immediately demolished by pointing out chemical impossibilities the head of the Cavendish Laboratory, Lawrence Bragg, told Watson and Crick to stop working on DNA structure and leave it to the King's College team. DNA was King's College's problem; Cambridge should focus on proteins.

Watson and Crick nominally obeyed this directive, officially abandoning DNA work. But they never really stopped thinking about the problem, never stopped collecting information, never stopped talking about possible structures. They remained alert to any new findings, any new insights, any pieces of information that might help solve the puzzle. And they maintained their contacts, particularly Wilkins, who increasingly shared information with them despite the awkwardness of going behind Franklin's back.

By late 1952 and early 1953, the situation was becoming increasingly tense. Pauling was known to be working on DNA structure, and rumors suggested he might be close to a solution. Franklin had obtained extraordinary X-ray crystallography data but was proceeding cautiously with interpretation, determined not to speculate beyond what her data definitively showed. Wilkins was frustrated by the lack of progress and the dysfunction of his relationship with Franklin. And Watson and Crick were champing at the bit, convinced they could solve the structure if only they had access to the right data and were allowed to build models freely.

The breakthrough, when it came, would emerge from a perfect storm of factors: excellent experimental data from Franklin, chemical principles from Chargaff, model-building expertise from Watson and Crick, and a moment of insight about how bases might pair. But it would also involve questionable ethics the sharing of Franklin's data with Watson and Crick without her knowledge or permission that would haunt the story and raise troubling questions about credit, collaboration, and the treatment of women in science.

Nine

Photo 51

"The Image That Changed Everything"

In May 1952, Rosalind Franklin positioned a carefully prepared fiber of DNA in the sample holder of her X-ray crystallography apparatus at King's College London. She had spent months perfecting her technique, learning to create DNA fibers that were unusually pure and well-aligned, with the long DNA molecules oriented parallel to each other like trees in a carefully planted forest. This alignment was crucial the better aligned the molecules, the sharper and more informative the X-ray diffraction pattern would be.

Franklin had discovered that DNA could exist in two different forms depending on humidity. At lower humidity, DNA adopted what she called the "A form," while at higher humidity it switched to the "B form." Most of her colleagues had been studying the A form, but Franklin suspected that the B form the form DNA would take under the wet conditions that prevailed inside living cells was more biologically relevant. She decided to focus her efforts on obtaining the best possible diffraction photographs of B-form

DNA.

The X-ray crystallography setup required patience and precision. The X-ray beam had to be precisely aligned, passing through the DNA sample and then onto photographic film that would record the diffraction pattern. Exposure times could run many hours, and any vibration or misalignment would blur the image. Franklin was meticulous about every detail, checking and rechecking her setup, ensuring that everything was perfect before beginning the long exposure.

When she developed the photographic plate after this particular exposure the photograph that would be labeled "Photo 51" in her lab notebook Franklin knew immediately that she had captured something spectacular. The image showed an extraordinarily clear X-shaped pattern, the signature of a helical structure, surrounded by a series of dark spots arranged in a systematic pattern. It was the clearest, most detailed diffraction pattern of DNA that anyone had ever obtained, and it was packed with structural information waiting to be decoded.

Reading the Photograph

To the untrained eye, Photo 51 might look like an abstract pattern, perhaps a piece of modern art dark spots and arcs arranged in a vaguely symmetric pattern on a lighter background. But to someone versed in X-ray crystallography, the photograph was a message written in the language of diffraction physics, telling a detailed story about DNA's three-dimensional structure.

The prominent X-shaped pattern dominating the photograph was the most obvious feature, and to a crystallography, it screamed "helix!" When X-rays pass through a helical structure, they produce this characteristic X-shaped diffraction pattern, with the angle and spacing of the X depending on the helix's pitch how tightly wound it is. The clarity and sharpness of the X in Photo 51 meant that DNA's helical structure was highly regular and well-defined.

The systematic arrangement of spots in the diffraction pattern contained additional information. The spacing between spots could be used to calculate the distance that repeated along the helix what crystallography call the "repeat distance." The relative intensities of different spots revealed information about what was located at different positions along and around the helix. Every detail of the photograph, properly analyzed, contained clues about DNA's architecture.

Franklin began the painstaking work of analyzing Photo 51, making careful measurements and applying the mathematical formalism of crystallography to extract structural information. She could determine that the helix made a complete turn every 34 angstroms (an angstrom is one ten-billionth of a meter). She could see that there were 10 base pairs per complete turn, meaning each base pair was separated by about 3.4 angstroms along the helix axis. She could deduce that the helical structure had two-fold symmetry, suggesting that it might consist of two intertwined strands running in opposite directions.

But Franklin was cautious about interpretation, perhaps overly so. She wanted to be absolutely certain about every conclusion, to nail down every detail before making public claims about DNA's structure. She knew that crystallographic data could sometimes be misleading, that apparent patterns might have alternative explanations. She continued collecting more photographs, more data, building up an overwhelming case for her structural conclusions rather than rushing to publish a preliminary model.

The Unauthorized Viewing

In January 1953, James Watson traveled from Cambridge to London to visit Maurice Wilkins at King's College. The exact circumstances of what happened next have been disputed, with different participants giving different accounts, but the essential facts are clear. Wilkins, apparently without thinking carefully about the implications, showed Watson a copy of Photo 51. Franklin had taken the photograph, but it had been given to Wilkins either as a professional courtesy or because of institutional sharing of data within the laboratory. Wilkins evidently felt entitled to show it to collaborators, though he did so without Franklin's knowledge or permission.

Watson's reaction upon seeing Photo 51 was visceral and immediate. Years later, he would write that his pulse began to race and his mouth went dry. He had seen enough crystallography to recognize the unmistakable signs of a helix, and the quality and clarity of Franklin's photograph far exceeded anything else he had seen. More importantly, the photograph provided specific measurements and constraints that would be invaluable for model building. Watson hurried back to Cambridge, excited to share what he had learned with Crick.

Shortly afterward, Crick obtained additional information from Franklin's work through a different channel. Max Perutz, Crick's supervisor at Cambridge, had received a copy of a research report that Franklin had submitted to a Medical Research Council committee, summarizing her crystallographic findings. The report contained detailed measurements and conclusions from her work, including precise dimensions of the helix. Perutz shared this report with Crick, apparently not realizing or not caring that Franklin might have considered this information confidential or at least not meant for competitors.

Armed with Franklin's data obtained without her knowledge through what most modern scientists would consider ethically dubious channels Watson and Crick now had the crucial pieces they needed to build an accurate model of DNA's structure. They knew it was a helix, they knew its dimensions, they knew key measurements about the spacing and arrangement of its components. Combined with Chargaff's rules about base pairing and their own chemical knowledge, they were poised to make the breakthrough.

The ethical questions raised by this episode are profound and have generated debate among historians and scientists for decades. Did Wilkins have the right to share Franklin's photograph with Watson? Should Perutz have given Crick access to Franklin's report? Were Watson and Crick obligated to tell Franklin they were using her data, or to invite her to collaborate? Different people have answered these questions differently, but there's no question that Franklin's data was used by Watson and Crick in building their model, and that this use occurred without her knowledge or consent.

Franklin herself would never know that her photograph had been shown to Watson at this crucial moment. She continued her careful analysis, unaware that the most important clue from her work was now in the hands of the Cambridge team. By the time she learned about Watson and Crick's model, it would already be published, and the moment for her to claim priority for the crucial experimental evidence would have passed.

Photo 51 represented years of Franklin's careful work, her technical expertise, her meticulous attention to detail. It was perhaps the most important single piece of experimental evidence in the discovery of DNA's structure. Yet Franklin's name would not appear on the famous paper announcing the double helix, and her contribution would be minimized or ignored in early accounts of the discovery. Only decades later would historians begin to recognize the full extent of her contribution and the problematic circumstances under which her data had been used.

Ten

The Double Helix

"Building Life with Wire and Sheet Metal"

In February 1953, with Franklin's crystallographic data fresh in their minds and the knowledge that Linus Pauling was apparently closing in on a DNA structure, Watson and Crick threw themselves into model building with renewed intensity. They obtained permission from Lawrence Bragg, their department head, to resume working on DNA structure. Bragg had changed his mind partly because of the Pauling threat if Pauling was going to solve the structure anyway, Cambridge might as well compete and partly because Perutz had convinced him that the crystallographic data was now good enough to make rapid progress possible.

Watson and Crick had learned from their earlier failures. Their disastrous 1951 attempt at building a DNA model had foundered on basic chemical errors and insufficient attention to the experimental constraints. This time, they would be more careful, more systematic, and better informed. They would build models that respected all the known constraints: the chemical composition determined by Levene decades earlier, the one-to-one base ratios discovered by Chargaff, the helical dimensions revealed by Franklin's crystallography, and the basic principles of chemical bonding that determined what structures were physically possible.

The Building Blocks

The machine shop at the Cavendish Laboratory fabricated metal plates cut to scale to represent the different bases adenine, guanine, cytosine, and thymine. Each plate was cut to the precise shape and size of the base molecule it represented, with holes drilled at the positions where atoms could form chemical bonds. Metal rods represented the bonds between atoms, and brass fittings connected the bases to the sugar-phosphate backbone. It was like a molecular construction set, allowing Watson and Crick to physically manipulate possible structures, testing what fit together and what didn't.

The constraints they worked with were numerous and specific. The sugar-phosphate backbone had to adopt a chemically reasonable configuration, with bond angles and lengths that matched known values. The bases had to stack roughly 3.4 angstroms apart, as Franklin's data indicated. The helix had to make a complete turn every 34 angstroms, meaning ten base pairs per turn. The structure had to have the two-fold symmetry that crystallography revealed. And somehow, it had to accommodate Chargaff's mysterious rules why did adenine always equal thymine in amount, and guanine always equal cytosine?

Watson and Crick initially tried models with the bases pointing outward, on the exterior of the helix, with the sugar-phosphate backbone running through the center. This seemed logical it would protect the information-carrying bases from chemical damage. But such models didn't work well. The bases pointing outward created an irregular, bumpy exterior that didn't match the crystallographic data, which suggested a smooth, regular structure.

Moreover, with bases on the outside, there was no obvious way to explain Chargaff's rules or how bases from one strand might interact with bases from another strand.

They tried three-stranded helices, thinking that three intertwined chains might provide stability. But the geometry didn't work out three strands created problems with bond angles and spacing that couldn't be resolved. They tried different arrangements of the sugar-phosphate backbone, different tilts and rotations of the bases, different everything. Model after model was built, examined, found wanting, and dismantled.

The Breakthrough

The crucial insight came from thinking carefully about base pairing and about the chemical structures of the bases themselves. Watson had been playing with the metal base templates on his desktop, trying different arrangements, when he realized that adenine and thymine, if properly aligned, had complementary shapes that would allow them to form two hydrogen bonds between them. Similarly, guanine and cytosine could form three hydrogen bonds when properly paired. The bonding wasn't random specific bases paired with specific partners, and the pairing was dictated by the chemical structures of the bases themselves.

Moreover, Watson noticed something remarkable about these pairings. Adenine and guanine were purines larger bases with two-ring chemical structures. Cytosine and thymine were pyrimidines smaller bases with single-ring structures. When a purine paired with a pyrimidine (adenine with thymine, or guanine with cytosine), the resulting base pair had essentially the same width. But pairing a purine with another purine would create a too-wide structure, while pairing a pyrimidine with another pyrimidine would be too narrow. Only purine-pyrimidine pairing maintained a constant width for the helix.

This explained Chargaff's rules instantly and beautifully. Adenine always equaled thymine because adenine on one strand always paired with thymine on the opposite strand. Guanine always equaled cytosine for the same reason. The one-to-one ratios weren't mysterious they were inevitable consequences of specific base pairing in a two-stranded structure. Every adenine in DNA was paired with a thymine, every guanine with a cytosine. The ratios had to be equal.

Watson rushed to share this insight with Crick, and they immediately began building a new model incorporating base pairing. The bases would point inward, toward the center of the helix, where they would pair with bases on the opposite strand through hydrogen bonding. The sugar-phosphate backbones would run along the outside, creating the smooth, regular exterior that crystallography indicated. The two strands would run in opposite directions anti parallel, in the technical jargon with one strand running 5' to 3' and the other running 3' to 5' (the numbers referring to carbon atoms in the sugar molecules).

As they built the model, everything clicked into place with satisfying precision. The base pairs stacked neatly 3.4 angstroms apart. The helix made one complete turn every 34 angstroms, accommodating exactly ten base pairs per turn. The structure had the required two-fold symmetry. The dimensions matched Franklin's measurements perfectly. The chemical bonds were all reasonable and stable. Chargaff's rules were explained. It worked. Every constraint was satisfied, every piece of evidence accommodated.

The Implications Reveal Themselves

As Watson and Crick stood back and looked at their completed model, they realized they had solved more than just a structural problem. The double helix structure immediately suggested how DNA could replicate how genetic information could be copied from one generation to the next. If the two strands separated, each strand could serve as a template for building a new complementary strand. Where there was an adenine on the template strand, a thymine would be added to the new strand. Where there was a guanine, a cytosine would be added. The base pairing rules ensured that each new double helix would be identical to the original.

This replication mechanism was so obvious, so elegant, so perfectly suggested by the structure itself, that Watson and Crick couldn't resist including a coy reference to it in their paper. In what might be the most famous understatement in scientific literature, they wrote: "It has not escaped our notice that the specific pairing we have postulated immediately suggests a possible copying mechanism for the genetic material."

The structure also explained how DNA could carry information. The sequence of bases along one strand could vary in any combination ATGC-GATCG (**Adenine-Thymine-Guanine-Cytosine repeated**) or any other arrangement of the four bases. With four bases and long strands, the number of possible sequences was astronomical, more than sufficient to encode the complex information needed to specify genes and direct development. The information was written in a four-letter alphabet, and the length of DNA molecules meant that this alphabet could spell out messages of unlimited complexity.

Watson and Crick invited colleagues from the Cavendish Laboratory to view their model. The response was enthusiastically positive. The structure was aesthetically beautiful, with the two strands twining around each other in a graceful spiral. It satisfied all the chemical and physical constraints. It explained the mysterious base ratios. And it immediately suggested how DNA could fulfill its genetic functions. The model was convincing enough that Watson and Crick felt confident submitting a paper to Nature, the prestigious British scientific journal that published short communications on important discoveries.

Racing to Publication

But there was still the problem of King's College. Watson and Crick's model was based partly on Franklin's data, obtained through questionable channels. They needed to handle the situation delicately to avoid accusations of stealing or scooping. The solution came through a coordinated publication strategy. Three papers would appear together in the same issue of Nature: Watson and Crick's paper presenting the structure, a paper from Franklin and her student Raymond Gosling presenting their crystallographic data, and a paper from Wilkins and his colleagues presenting their supporting evidence.

This arrangement was meant to provide everyone with appropriate credit Watson and Crick for proposing the structure, Franklin for the crucial crystallographic evidence, and Wilkins for his supporting work. But the reality was more complicated. Franklin's paper, written quickly to accompany Watson and Crick's, presented her data as supporting evidence for the structure rather than as the independent work that had actually preceded and enabled the structure's determination. She had not been consulted about using her data; she learned about the double helix only shortly before publication.

Watson and Crick's paper appeared in Nature on April 25, 1953. It was astonishingly brief only 900 words long, barely more than two pages. There was a simple diagram showing the base pairing and the overall helical structure. The writing was clear and confident, presenting the structure as an established fact rather than a tentative hypothesis. The paper acknowledged Franklin's and Wilkins's experimental work in rather perfunctory terms, not mentioning the crucial role that Photo 51 had played in constraining the model.

The scientific community's response was immediate and overwhelmingly positive. The structure made too much sense, explained too much, to be wrong. Within months, additional evidence confirmed the model's correctness. The double helix became established as scientific fact with remarkable speed, and Watson and Crick became scientific celebrities. Molecular biology had its foundational structure, and a new era of biological research based on understanding DNA at the atomic level had begun.

Eleven

Controversy and Recognition

"The Shadows Behind the Helix"

The announcement of DNA's structure in 1953 should have been an unalloyed triumph, a moment when science revealed one of nature's most fundamental secrets and all the contributors received their due recognition. Instead, it became a story marked by controversy, hurt feelings, missed credit, and persistent questions about ethics and fairness that continue to trouble the scientific community seventy years later.

The immediate aftermath seemed positive enough. Watson, Crick, Wilkins, and Franklin all published papers in the same issue of Nature, a coordinated release that seemed to suggest collegial collaboration. The scientific community accepted the double helix structure enthusiastically, recognizing it as a major breakthrough. Laboratories around the world began reorienting their research to take advantage of this new understanding of DNA. The structure was reproduced in textbooks, taught to students, and became one of the most recognizable icons in all of science.

But beneath this surface success, serious problems lurked. Franklin's crucial contribution the crystallographic data that had constrained and validated the model was underappreciated and sometimes actively minimized. Watson and Crick's use of her data without her knowledge or permission raised ethical questions that weren't publicly discussed at the time but would later generate intense debate. And as recognition and honors began to flow toward the discoverers of DNA's structure, the distribution of credit would prove deeply problematic.

The Franklin Question

Rosalind Franklin never publicly complained about how her work had been used. After the publication of the double helix papers, she moved to a different institution and shifted her research to viruses, where she made important contributions studying the structure of tobacco mosaic virus and other viral particles. She remained cordial with Wilkins, despite their difficult history at King's College, and apparently bore no grudge against Watson and Crick, though she couldn't have known the full story of how her data had reached them.

In 1956, Franklin began experiencing health problems. She was diagnosed with ovarian cancer, a disease that in the 1950s was usually fatal. She continued working as long as she could, maintaining her research program even as her health declined. She died on April 16, 1958, at the age of thirty-seven, just five years after the double helix structure was published. Her death was almost certainly caused by her extensive exposure to X-rays during her crystallographic work radiation safety standards in the 1950s were minimal, and Franklin had worked with X-ray equipment for years without adequate shielding. She sacrificed her health, and ultimately her life, in pursuit of scientific knowledge.

Franklin's early death had profound consequences for how her contribution was remembered and recognized. Nobel Prizes are not awarded posthumously, and by the time the Nobel Prize in Physiology or Medicine was awarded in 1962 for the discovery of DNA's structure, Franklin was not alive to receive it. The prize went to Watson, Crick, and Wilkins three men. Franklin's name was not mentioned in the Nobel lectures except in passing. Her crucial contribution seemed to have been forgotten or at least minimized.

Watson's 1968 book "The Double Helix," a bestselling memoir of the discovery, made the situation worse. Watson portrayed Franklin as difficult, unfeminine, and obstructive what he infamously called her "Rosy," a nickname she never used and reportedly disliked. He depicted her as someone who didn't understand her own data and needed Watson and Crick to interpret it properly. The portrait was dismissive and condescending, presenting Franklin as an obstacle to be overcome rather than a crucial contributor to be acknowledged.

The book generated controversy immediately upon publication. Many scientists, including Crick, felt that Watson had distorted events and treated Franklin unfairly. Franklin's friends and colleagues were outraged by the portrayal. But the damage was done Watson's book was widely read, and his version of events became the standard narrative for a generation of scientists and students. The brilliant, meticulous scientist who had produced Photo 51 became "Rosy," the difficult woman who didn't appreciate her own work.

The Rehabilitation

In the decades following her death, historians of science began investigating Franklin's actual contributions more carefully, examining her notebooks, reading her papers, and interviewing people who had worked with her. What emerged was a very different picture from Watson's portrait. Franklin had not been oblivious to the significance of her data her notebooks show that she had deduced most of the key features of DNA's structure before Watson and Crick published their model. She knew it was a helix, she had calculated its dimensions accurately, she had recognized the anti parallel nature of the strands. She had been cautious about publishing because she wanted to be absolutely certain, not because she didn't understand what she was seeing.

The question of whether Franklin would have solved the complete structure herself if given more time remains counterfactual and ultimately unanswerable. She might have, or she might have required the additional insight about base pairing that Watson provided. But there's no question that her crystallographic work was essential to the discovery. Watson and Crick built their model to fit Franklin's data without Photo 51 and the measurements it provided, they couldn't have determined the correct structure with confidence.

Biographies published in the 1970s and later began giving Franklin her due credit. "Rosalind Franklin: The Dark Lady of DNA" by Brenda Maddox, published in 2002, provided a comprehensive and sympathetic account of Franklin's life and work. The scientific community increasingly recognized that Franklin deserved to be remembered as a co-discoverer of DNA's structure, even if the Nobel Prize rules prevented her from sharing the official recognition.

Modern accounts of the discovery typically present Franklin as one of the four key figures, alongside Watson, Crick, and Wilkins. Her name appears on commemorative plaques and in historical summaries. The European Space Agency named a Mars rover after her. She has become an icon for women in science, a reminder of how female scientists' contributions have often been overlooked or minimized, and an inspiration for current generations pursuing scientific careers despite obstacles and discrimination.

Yet the controversy persists. Was Franklin robbed of credit she deserved?

Were Watson and Crick scientific thieves who stole her data and claimed her discoveries as their own? Or were they simply working in a competitive environment where data sharing was informal and attribution standards were different from modern norms? Different historians and scientists have reached different conclusions, and the debate shows no signs of ending.

Other Overlooked Contributors

Franklin wasn't the only important contributor whose recognition fell short of their contribution. Oswald Avery, whose transformation experiments had proven that DNA was the genetic material, never received the Nobel Prize despite being nominated multiple times. He died in 1955, three years before the prize was awarded to another DNA-related discovery, and by 1962 when the prize was awarded for DNA structure, he had been dead for seven years. Many scientists believe Avery should have won the prize during his lifetime and that the failure to award it to him was one of the Nobel committee's most serious oversights.

Erwin Chargaff, whose rules about base pairing were essential to deducing DNA's structure, felt slighted that he wasn't included in the 1962 Nobel Prize. His rules had revealed the base pairing that was central to the double helix structure, yet he received no official recognition beyond citations in scientific papers. Chargaff became increasingly bitter about this perceived slight, and his later writings about the discovery have a sour tone, criticizing Watson and Crick for not properly acknowledging his contribution.

Raymond Gosling, Franklin's graduate student who had actually taken Photo 51 under Franklin's direction, is rarely mentioned in historical accounts. He did important work but was overshadowed by his more famous supervisor. Linus Pauling, despite being arguably the world's leading structural chemist, had failed to solve DNA's structure and had in fact published an incorrect model shortly before Watson and Crick announced the double helix. His failure, caused partly by insufficient access to crystallographic data and partly by uncharacteristic chemical errors, meant he was left out of the DNA structure story despite his enormous contributions to structural biology more broadly.

The 1962 Nobel Prize

When the Nobel Prize in Physiology or Medicine was announced in October 1962, the award went to James Watson, Francis Crick, and Maurice Wilkins "for their discoveries concerning the molecular structure of nucleic acids and its significance for information transfer in living material." The choice of these three recipients reflected the Nobel committee's judgment about who had made the most important contributions, but it also reflected the constraints of Nobel rules only three living recipients could share a prize, and posthumous awards were not permitted.

If Franklin had lived, would she have shared the prize? Many people believe she should have, that a four-way split among Watson, Crick, Wilkins, and Franklin would have been more just. But the Nobel Prize's three-recipient limit would have made that impossible. Perhaps Wilkins would have been left out to make room for Franklin, or perhaps the committee would have found some other way to draw lines between contributors. We'll never know.

Watson, Crick, and Wilkins each gave Nobel lectures explaining their work. Watson and Crick's lectures focused on the structure itself and its implications. Wilkins's lecture emphasized the crystallographic evidence. Franklin's contribution was mentioned, but only briefly and in passing, as one source of experimental data among others. The Nobel lectures, like the prize itself, seemed to suggest that Watson, Crick, and Wilkins were the sole or at least primary discoverers, with others playing supporting roles.

Lessons and Lingering Questions

The controversy surrounding DNA's discovery raises important questions that remain relevant today. How should credit be distributed in collaborative science? What obligations do scientists have to acknowledge data and insights they've received from others? How do we ensure that all contributors, regardless of gender or status, receive appropriate recognition? When competition drives scientists to questionable ethical behavior using others' data without permission, minimizing others' contributions how should the scientific community respond?

Modern science has tried to address some of these issues through more formal systems of attribution and collaboration. Multi-author papers with clearly defined contributions have become standard. Data sharing is more regulated, with explicit agreements about usage and credit. Ethics review and oversight have become more rigorous. Efforts to increase diversity in science and combat discrimination have made the field more aware of how bias can affect who receives credit and recognition.

But problems persist. Women and minorities in science still face obstacles to advancement and recognition. Credit disputes still arise in competitive fields where multiple groups work on similar problems. The pressure to publish first and claim priority still drives some scientists to ethically questionable behavior. The DNA structure story, with all its triumphs and troubles, remains relevant as both inspiration and cautionary tale.

Perhaps the most important lesson is that scientific discoveries rarely have simple, clean narratives with clear heroes and villains. The discovery of DNA's structure involved brilliant insights, hard work, technical expertise, creative thinking, and yes, some ethically questionable data sharing. It involved both men and women, junior and senior scientists, multiple institutions and nationalities. It built on decades of previous work by chemists, geneticists, and crystallographers. Trying to assign sole credit to any individual or even any small group inevitably does injustice to the complex, collaborative nature of scientific progress.

The double helix structure stands as one of humanity's great intellectual achievements, a discovery that revolutionized biology and continues to transform medicine, agriculture, and our understanding of life itself. We can celebrate that achievement while also honestly acknowledging the messy human story behind it the competition and collaboration, the credit and controversy, the brilliance and the flaws. Both aspects of the story deserve to be remembered and learned from.

Twelve

The Central Dogma

"From DNA to Proteins - The Flow of Information"

With DNA's structure revealed, molecular biologists faced a new set of questions. The double helix explained how genetic information could be stored in the sequence of bases and how DNA could replicate by strand separation and complementary base pairing. But a crucial mystery remained: how did the information encoded in DNA actually direct the synthesis of proteins, the molecules that did most of the actual work in cells? How did the four-letter alphabet of DNA bases get translated into the twenty-letter alphabet of amino acids that made up proteins?

Francis Crick, never content with solving just one problem when several remained, began thinking systematically about information flow in biological systems. In 1958, he proposed what he called the "Central Dogma" of molecular biology, a framework for understanding how genetic information moves within cells. The central dogma stated that information flows in one direction: from DNA to RNA to protein. DNA serves as the master copy of genetic information, stored safely in the nucleus. RNA serves as a temporary working copy, carrying information from DNA to the protein-synthesis machinery. And proteins serve as the end products, the molecules that actually perform cellular functions.

The central dogma wasn't just an abstract principle it reflected specific molecular processes that scientists were beginning to understand. The first step, transcription, involved copying DNA's sequence into RNA. An enzyme called RNA polymerase would bind to DNA, separate the two strands locally, and synthesize an RNA molecule using one DNA strand as a template. The RNA molecule was complementary to the DNA template strand, built using the same base-pairing rules that governed DNA replication, with one exception: RNA used the base uracil (U) instead of thymine (T). Where DNA had thymine, RNA had uracil, but U paired with adenine just as T did, so the information transfer was faithful.

The second step, translation, was more complex. RNA molecules called messenger RNA, or mRNA, would carry genetic information from the nucleus to ribosomes, complex molecular machines that synthesized proteins. At the ribosome, the mRNA sequence would be read in groups of three bases called codons, and each codon would specify a particular amino acid to be added to the growing protein chain. Transfer RNA molecules, or tRNA, would bring the appropriate amino acids to the ribosome, matching their anticodon sequences to the codons on the mRNA. Through this process, the four-letter language of nucleic acids was translated into the twenty-letter language of proteins.

This information flow DNA to RNA to protein became one of the fundamental principles organizing molecular biology. It explained how genes worked at the molecular level: genes were stretches of DNA that encoded RNA molecules, which in turn directed protein synthesis. Mutations in DNA changed the sequence of bases, which changed the sequence of amino acids in proteins, which altered protein function and thereby affected the organism's characteristics. The molecular basis of heredity, of genetic variation, of evolution itself, could all be understood through this framework.

The Genetic Code Problem

But understanding the overall information flow didn't solve all the mysteries. A crucial puzzle remained: what was the genetic code? How did specific sequences of DNA bases correspond to specific amino acids? With four bases and twenty amino acids, the code couldn't be one-to-one-four bases could only specify four amino acids. Even if bases were read in pairs, that would only give sixteen possible combinations (4 × 4), still not enough for twenty amino acids. The code had to use at least three bases per amino acid, giving sixty-four possible combinations (4 × 4 × 4), which would be more than sufficient.

Simple arithmetic suggested a triplet code, but which triplets corresponded to which amino acids? Were there commas or spacing between codons, or was the code read continuously? Did some codons serve as punctuation, marking where one gene ended and another began? Was the code the same in all organisms, or did different species use different codes? These questions demanded experimental answers, and getting those answers would require developing entirely new techniques for studying protein synthesis outside of living cells.

The breakthrough came from an unlikely source and an almost absurdly simple experiment. In 1961, Marshall Nirenberg and Heinrich Matthaei at the National Institutes of Health in Bethesda, Maryland, succeeded in creating a cell-free system that could synthesize proteins. They ground up bacterial cells, extracted the cellular machinery needed for protein synthesis ribosomes, transfer RNAs, amino acids, and all the necessary enzymes and mixed these components together in a test tube. When they added messenger

RNA to this mixture, proteins were synthesized. They had created artificial life in a test tube, or at least the protein-making part of life.

Nirenberg and Matthaei then did something ingenious. Instead of using natural mRNA extracted from cells, which would contain complex mixtures of different sequences, they used synthetic RNA polymers made in the laboratory. Their first synthetic RNA was polyuridylic acid, or poly-U-a long chain consisting entirely of uracil bases, the RNA equivalent of a string of U's: UUUUUUUUU. When they added poly-U to their cell-free system along with a mixture of amino acids, one of which was radioactively labeled, protein synthesis occurred. Even more remarkably, the resulting protein consisted entirely of one amino acid: phenylalanine.

The conclusion was inescapable. The codon UUU specified the amino acid phenylalanine. It was the first crack in the code, the first definitive assignment of a specific nucleotide sequence to a specific amino acid. Nirenberg announced the finding at a small session at an international congress in Moscow in August 1961, expecting little attention. Instead, Francis Crick, who was in the audience, immediately recognized the breakthrough's significance. He insisted that Nirenberg present his findings to the entire congress, and suddenly Nirenberg went from being an obscure researcher to being a major figure in molecular biology.

Breaking the Code

The poly-U experiment opened the floodgates. If synthetic RNA polymers could be used to determine codon assignments, then the entire genetic code could potentially be cracked through systematic experiments. Nirenberg's laboratory, joined by several other research groups, began a race to decode all sixty-four possible triplet combinations. They synthesized RNA polymers with different base compositions poly-A (all adenines), poly-C (all cytosines), polymers with random mixtures of two or three different bases, and eventually polymers with specific defined sequences.

Each synthetic RNA was added to a cell-free protein synthesis system, and the resulting proteins were analyzed to determine which amino acids had been incorporated. Some experiments were straightforward: poly-A directed synthesis of polylysine, indicating that AAA coded for lysine. Poly-C produced polyproline, so CCC coded for proline. But experiments with mixed polymers were more complex, requiring statistical analysis to tease out which codons specified which amino acids.

Har Gobind Khorana at the University of Wisconsin developed a complementary approach, using chemical synthesis to create RNA molecules with precisely defined sequences. Instead of random polymers, Khorana could make RNA chains with specific repeating patterns UCUCUCUCUC or AGAGAGAGAG and determine which amino acids these sequences specified. His meticulous chemical synthesis work provided crucial data that helped nail down codon assignments with certainty.

By 1965, just four years after the first codon assignment, the complete genetic code had been deciphered. All sixty-four possible triplet codons had been assigned meanings. Most of the codons specified one of the twenty standard amino acids. Three codons UAA, UAG, and UGA didn't code for any amino acid but instead served as stop signals, telling the ribosome that protein synthesis should terminate. One codon, AUG, had a dual function: it coded for the amino acid methionine, and it also served as a start signal, marking where translation should begin.

The code proved to be redundant, with multiple codons often specifying the same amino acid. For instance, leucine was specified by six different codons: UUA, UUG, CUU, CUC, CUA, and CUG. This redundancy, called degeneracy, provided a buffer against mutations. If a mutation changed the third position of a leucine codon from UUA to UUG, the amino acid sequence wouldn't change both codons still specified leucine. The genetic code had built-in error tolerance.

The Universal Code

Perhaps the most remarkable discovery about the genetic code was its universality. With very few exceptions, all living organisms used the same code. The codon UUU specified phenylalanine in bacteria, in yeast, in plants, in insects, in humans. GCA specified alanine in all organisms. The entire sixty-four-codon dictionary was essentially identical across all of life.

This universality had profound implications. It suggested that the genetic code had been established very early in the history of life, probably in the last universal common ancestor of all living things, and had been maintained essentially unchanged for billions of years of evolution. It demonstrated the fundamental unity of life at the molecular level all organisms spoke the same genetic language, used the same molecular machinery to read that language, and built their proteins using the same amino acid alphabet.

The universality of the genetic code also made genetic engineering possible. If all organisms used the same code, then DNA from one organism could be expressed in another organism. A human gene could be inserted into a bacterium, and the bacterium would read that gene correctly and produce the human protein. This realization, which came in the late 1960s and early 1970s, would launch the biotechnology revolution and transform medicine, agriculture, and industry.

For their work in cracking the genetic code, Nirenberg, Khorana, and Robert Holley (who had determined the structure of transfer RNA) shared the 1968 Nobel Prize in Physiology or Medicine. The prize recognized not just the technical achievement of assigning codons to amino acids, but the fundamental insight this work provided into how life stores and uses genetic information. The genetic code was the Rosetta Stone that allowed translation between the language of nucleic acids and the language of proteins, and understanding that code was essential to understanding molecular biology.

Thirteen

DNA Replication Revealed

"The Copy Machine of Life"

Watson and Crick's famous understatement in their 1953 paper "It has not escaped our notice that the specific pairing we have postulated immediately suggests a possible copying mechanism for the genetic material" had pointed toward how DNA might replicate, but pointing toward a mechanism and proving that mechanism was correct were two very different things. The double helix structure suggested that DNA replication might work through strand separation and complementary base pairing, but this remained a hypothesis that needed experimental testing.

The replication problem was profound and multifaceted. How did the two strands of the double helix separate, given that they were intertwined like the strands of a rope? How was the base sequence of the original strand copied onto a new strand with sufficient accuracy to avoid introducing errors? What molecular machinery performed the actual synthesis of new DNA strands? And most fundamentally, did replication really work through the semi-conservative mechanism suggested by the structure, where each original strand served as a template for a new strand, or might it work through some other mechanism?

Three possible replication mechanisms could be imagined. Conservative replication would keep the original DNA double helix intact and create an entirely new double helix using the original as a template like making a photocopy where the original remains unchanged and a complete copy is produced. Dispersive replication would somehow break up the original DNA and distribute fragments of old and new DNA throughout both daughter molecules. Semi-conservative replication, suggested by the double helix structure, would separate the two strands of the original DNA, and each strand would serve as a template for synthesizing a new complementary strand, resulting in two DNA molecules each containing one old strand and one new strand.

The Most Beautiful Experiment

In 1958, two young scientists at the California Institute of Technology, Matthew Meselson and Franklin Stahl, published what would become known as "the most beautiful experiment in biology"—a definitive test of the three replication mechanisms that was so elegant in design and so unambiguous in results that it immediately settled the question.

Meselson and Stahl's experiment exploited the existence of heavy nitrogen, an isotope of nitrogen that is slightly heavier than the normal nitrogen-14 that predominates in nature. Nitrogen-15 isn't radioactive it's stable and chemically identical to nitrogen-14 but its extra neutron makes molecules containing it denser and heavier. DNA contains nitrogen in its bases, so DNA made with nitrogen-15 would be denser than DNA made with normal nitrogen-14. This density difference, though tiny, could be detected using a technique called cesium chloride density gradient centrifugation, which separated molecules based on their densities.

The experimental design was straightforward in concept but technically demanding in execution. Meselson and Stahl grew bacteria for many generations in medium containing nitrogen-15, ensuring that all the nitrogen in the bacteria's DNA came from the heavy isotope. This produced DNA that was fully "heavy," with nitrogen-15 in both strands. Then they abruptly switched the bacteria to medium containing only normal nitrogen-14 and allowed the bacteria to continue growing and dividing. Any new DNA synthesized after the switch would incorporate light nitrogen-14, while the original DNA contained heavy nitrogen-15.

The question was: what would happen to the DNA density as bacteria replicated their DNA in the presence of light nitrogen? The answer would depend on which replication mechanism was correct. Conservative replication would produce some DNA molecules that were fully heavy (the original molecules, preserved intact) and some that were fully light (the newly synthesized molecules). Dispersive replication would produce DNA of intermediate density, with fragments of heavy and light DNA mixed together. Semi-conservative replication would produce DNA of intermediate density in the first generation (each molecule containing one heavy strand from the original DNA and one light strand newly synthesized), and a mixture of intermediate and fully light DNA in the second generation.

Meselson and Stahl extracted DNA from bacteria at various times after the switch to light nitrogen medium and analyzed the DNA's density using cesium chloride centrifugation. The results were beautifully clear and unambiguous. After one generation of growth in light medium, all the DNA had intermediate density not fully heavy, not fully light, but exactly halfway between. After two generations, half the DNA had intermediate density and half had fully light density. The pattern perfectly matched the predictions of semi-conservative replication.

The conclusion was definitive: DNA replicated through strand separation and complementary base pairing, exactly as Watson and Crick had suggested. Each strand of the parental DNA served as a template for synthesizing a new strand. The two daughter DNA molecules each contained one strand from the parent and one newly synthesized strand. The base-pairing rules ensured that the new strands were exact complements of the template strands, and therefore the daughter DNA molecules were identical to the parent.

The Replication Machinery

Proving that replication was semi-conservative answered the fundamental question about the mechanism, but it raised new questions about the molecular machinery that actually performed DNA synthesis. How did the cell separate the two strands of the double helix, which were held together by hydrogen bonds and intertwined around each other? What enzyme or enzymes synthesized new DNA strands? How did the synthesis machinery achieve the remarkable accuracy needed to copy billions of base pairs with only rare errors?

The enzyme responsible for DNA synthesis, DNA polymerase, was discovered by Arthur Kornberg at Washington University in St. Louis in the late 1950s. Kornberg had been searching for the DNA-synthesizing enzyme by testing cell extracts for the ability to incorporate radioactive nucleotides into DNA. After years of painstaking work purifying and characterizing the enzyme, Kornberg isolated DNA polymerase and demonstrated that it could synthesize DNA in a test tube using a DNA template and a supply of nucleotide building blocks.

DNA polymerase turned out to be a remarkably sophisticated molecular machine. It could read the template DNA strand and add the complementary nucleotide with exquisite specificity adding a T opposite an A, a G opposite a

C, and so on. It had proofreading capability, meaning it could detect when it had made an error and correct that error before continuing synthesis. This proofreading function was essential for maintaining genetic accuracy, ensuring that DNA replication introduced errors in only about one base per billion bases copied.

Further research revealed that DNA replication was even more complex than initially appreciated. The two strands of DNA ran in opposite directions—one strand ran 5' to 3' while the other ran 3' to 5'. DNA polymerase could only synthesize DNA in the 5' to 3' direction, creating a problem for replicating the strand running 3' to 5'. The solution, discovered in the late 1960s, was that one strand (the "leading strand") was synthesized continuously in the same direction that the replication fork moved, while the other strand (the "lagging strand") was synthesized discontinuously in short fragments that were later joined together.

Additional proteins were required for DNA replication beyond just DNA polymerase. Helicase enzymes separated the two strands of the double helix, breaking the hydrogen bonds between bases and unwinding the helix. Single-strand binding proteins kept the separated strands from re-annealing prematurely. Primase synthesized short RNA primers that DNA polymerase needed to start synthesis. Ligase enzymes joined the discontinuous fragments on the lagging strand together into a continuous strand. The complete replication machinery, called the replisome, was a complex assembly of multiple proteins working in coordination.

The Accuracy Imperative

The accuracy of DNA replication was crucial for life. The human genome contains about three billion base pairs, and every time a cell divided, all three billion base pairs had to be copied. If the error rate were too high say, one error per thousand bases each round of replication would introduce millions of errors into the genome, leading to dysfunctional genes, cellular chaos, and death. The actual error rate, thanks to DNA polymerase's proofreading capability and additional DNA repair mechanisms, was around one error per billion bases, low enough that most cell divisions occurred without any errors at all in critical genes.

When errors did occur, they became mutations permanent changes in the DNA sequence that could be passed to offspring. Most mutations were harmless, occurring in regions of DNA that didn't code for important functions. Some mutations were harmful, disrupting genes and causing disease or dysfunction. And very rarely, mutations were beneficial, providing some advantage that natural selection could favor. These beneficial mutations were the raw material for evolution, the source of genetic variation that allowed organisms to adapt to changing environments over generations.

Understanding DNA replication provided the molecular foundation for understanding heredity, variation, and evolution. Traits were passed from parents to offspring because DNA was accurately copied during reproduction. Offspring resembled their parents because the DNA sequence, and therefore the genes it encoded, remained largely unchanged. Yet offspring weren't identical to parents because occasional replication errors introduced mutations that created variation. Natural selection acted on this variation, favoring individuals whose particular DNA sequences gave them advantages in survival and reproduction.

The replication mechanism also explained the stability of species. DNA's remarkable replication accuracy meant that successful genetic programs could be maintained essentially unchanged over countless generations. The genes that worked well genes that produced functional proteins, genes that regulated development correctly, genes that allowed effective responses to environmental challenges were faithfully copied and preserved. Species could maintain their characteristics because DNA replication preserved those characteristics at the molecular level.

By the early 1960s, molecular biologists had answered the fundamental questions about DNA that had driven their field since the discovery of the double helix. They understood how DNA stored information, how it replicated, how it directed protein synthesis, and how the genetic code translated nucleotide sequences into amino acid sequences. The basic framework of molecular biology was complete, the central dogma validated, and the molecular basis of heredity revealed.

But understanding DNA's structure and function was just the beginning. The next challenges would involve actually reading DNA sequences, manipulating them in the laboratory, and ultimately using that capability to transform medicine, agriculture, and biotechnology. The revolution that DNA's discovery had started would accelerate dramatically in the coming decades, as scientists developed tools to read, write, edit, and engineer the genetic code itself.

Fourteen

Reading DNA

"From Genes to Genomes"

For two decades after the discovery of DNA's structure, molecular biologists understood the principles of how DNA stored and transmitted genetic information, but they had no practical way to read the actual sequences of bases in DNA molecules. They knew that genes were written in a four-letter alphabet, that different genes had different sequences, and that those sequences determined the amino acid sequences of proteins. But determining what any particular gene's sequence actually was remained impossible with the available technology.

The problem was fundamental. DNA molecules were extraordinarily long, containing millions or billions of bases, and the bases were chemically similar enough that distinguishing them in sequence was technically challenging. It was like trying to read a book by examining the molecular structure of its ink, or trying to determine what a computer program did by chemically analyzing the silicon in its memory chips. The information was there, encoded in the physical structure, but extracting that information required new approaches and new technologies.

The breakthrough came in 1977 through two different sequencing methods developed independently by Frederick Sanger in Cambridge, England, and Walter Gilbert and Allan Maxam at Harvard University. Both methods allowed researchers to determine the exact sequence of bases in DNA molecules, though they used different chemical principles. Sanger's method, called the chain termination method or dideoxy sequencing, proved to be more practical and became the standard technique used in laboratories around the world.

Sanger's Method

Fred Sanger was already a legendary figure in biochemistry when he turned his attention to DNA sequencing. He had won the Nobel Prize in Chemistry in 1958 for determining the amino acid sequence of insulin, the first protein ever sequenced. That work had taken him nearly a decade of painstaking effort using chemical degradation methods that broke proteins into smaller fragments, analyzed those fragments, and gradually pieced together the complete sequence like solving a complex jigsaw puzzle. Sanger's insulin work had proven that proteins had defined sequences and that those sequences could be determined, but the methods were slow and labor-intensive.

For DNA sequencing, Sanger developed a completely different approach. Instead of breaking DNA apart and analyzing the fragments, his method synthesized new DNA strands and then analyzed the sizes of those newly synthesized molecules. The technique exploited a clever chemical trick: using modified nucleotides called dideoxynucleotides that could be incorporated into growing DNA chains but, once incorporated, prevented any further elongation. The DNA synthesis would terminate at the point where the dideoxynucleotide was added.

In practice, Sanger would set up four separate DNA synthesis reactions, each containing normal nucleotides plus a small amount of one type of dideoxynucleotide ddA, ddG, ddC, or ddT. In the reaction with ddA, DNA synthesis would randomly terminate at positions where A normally appeared, producing a collection of DNA fragments of various lengths, each ending in A. Similarly for the other three reactions. By running these four reactions in parallel and separating the resulting DNA fragments by size using gel electrophoresis, Sanger could determine the sequence. The smallest fragment indicated where the first base was, the next smallest where the second base was, and so on, reading up the gel to determine the complete sequence.

Sanger's first major demonstration of his sequencing method was determining the complete sequence of the bacteriophage φX174 genome in 1977. This viral genome contained 5,386 bases—tiny by modern standards but an extraordinary achievement at the time. It was the first complete genome of any organism ever sequenced, and it revealed details about viral genes, regulatory sequences, and genome organization that had been invisible before. The sequence showed where genes started and stopped, how genes overlapped in some cases to pack maximum information into minimum space, and how the virus controlled its life cycle at the molecular level.

The impact was immediate and profound. For the first time, scientists could read the exact genetic instructions that made an organism what it was. They could compare sequences between different organisms and see precisely how genes had changed during evolution. They could identify mutations causing genetic diseases by comparing normal and mutant gene sequences. The genetic code, which had been deciphered in terms of which codons specified which amino acids, could now be read in actual genes.

The Automation Revolution

Sanger's method, while revolutionary, was initially labor-intensive and time-consuming. Each sequence determination required multiple reactions, careful preparation of reagents, running gels, and reading the sequences from autobiographers or other detection methods. Sequencing even a single gene required weeks of work, and sequencing anything longer seemed impractical.Sanger's method, while revolutionary, was initially labor-intensive and time-consuming. Each sequence determination required multiple reactions, careful preparation of reagents, running gels, and reading the sequences from autobiographers or other detection methods. Sequencing even a single gene required weeks of work, and sequencing anything longer seemed impractical.

The solution came through automation. In the 1980s, scientists developed automated DNA sequencing machines that could perform Sanger sequencing much faster and more accurately than humans working manually. The key innovation was using fluorescent dyes instead of radioactivity to label the DNA fragments, with each of the four bases labeled with a different colored dye. This allowed all four sequencing reactions to be run together in a single tube rather than four separate reactions, and the results could be read automatically by detecting the fluorescent colors as DNA fragments passed through a laser beam during gel electrophoresis.

Automated sequencing transformed DNA sequencing from a specialized technique practiced in a few laboratories into a routine tool available to any molecular biology lab. The cost per base sequenced dropped dramatically, from dollars to cents to fractions of pennies. The speed increased from hundreds of bases per day to thousands to millions. Projects that would have been completely impossible became merely difficult, then became routine.

The first major beneficiary of automated sequencing was the Human Genome Project, an international scientific collaboration launched in 1990 with the audacious goal of sequencing the entire human genome all three billion base pairs of DNA that make up the genetic instructions for a human being. When first proposed in the late 1980s, the project seemed almost fantastically ambitious. At the sequencing speeds available then, determining the human genome sequence would take centuries and cost astronomical sums. But advocates believed that technological improvements during the project would make it feasible, and they were right.

The Human Genome Project

The Human Genome Project was biology's equivalent of the Apollo moon landing a massive, coordinated effort involving thousands of scientists across multiple countries, driven by a clear goal and a defined timeline. The project officially began in 1990 with a projected completion date of 2005 and an estimated cost of three billion dollars, roughly one dollar per base pair. It represented the largest biological research project ever attempted, dwarfing anything that had come before in scope, cost, and complexity.

The project faced enormous technical challenges. Simply growing and maintaining the bacterial cultures needed to clone human DNA fragments for sequencing required industrial-scale operations. Developing software to assemble millions of short sequence reads into the correct order for chromosomes that might contain hundreds of millions of bases demanded sophisticated computational algorithms. Ensuring quality control across multiple sequencing centers in different countries required careful coordination and standardization.

The project also faced competition from private industry. In 1998, Craig Venter announced that his company, Celera Genomics, would sequence the human genome using a different, faster approach and would complete the work before the public consortium. This sparked a dramatic race between the public and private efforts, with both sides working at breakneck pace and publicly competing for primacy. The competition was sometimes bitter, but it also accelerated progress, driving both groups to work faster and develop better methods.

In February 2001, both groups published draft sequences of the human genome simultaneously in the journals Nature and Science, with coordinated press conferences announcing that the human genetic code had been cracked. It was a moment of enormous scientific triumph, celebrated globally as one of humanity's great intellectual achievements. The draft sequences weren't complete they contained gaps and uncertainties but they represented roughly 90% coverage of the genome and provided unprecedented insight into human genetics.

The "finished" human genome sequence, with higher accuracy and better coverage, was declared complete in April 2003, almost exactly fifty years after Watson and Crick's announcement of the double helix structure. The final cost of the public effort was about 2.7 billion dollars, roughly on budget, and the project had finished two years ahead of its original schedule.

What the Genome Revealed

The human genome sequence revealed surprises that overturned many assumptions about human genetics. Scientists had expected to find around 100,000 genes, based on estimates of the number of different proteins humans make and the complexity of human biology. Instead, the genome contained only about 20,000 to 25,000 protein-coding genes, not many more than simpler organisms like roundworms. This was humbling humans weren't genetically complex in the way scientists had assumed.

The genome showed that only about 1.5% of human DNA actually coded for proteins. The remaining 98.5% consisted of regulatory sequences, structural elements, repetitive sequences, and large stretches whose function remained unclear. Some scientists initially dismissed this as "junk DNA," but subsequent research revealed that much of it served important regulatory roles, controlling when and where genes were expressed.

The genome revealed the history of human evolution written in the DNA sequence. Ancient viral sequences that had inserted themselves into our ancestors' genomes millions of years ago remained recognizable. Pseudo genes broken copies of functional genes that had accumulated mutations and become nonfunctional told stories of evolutionary experiments that hadn't worked out. Comparisons between human and chimpanzee genomes (the chimp genome was sequenced in 2005) showed that we differed by only about 1.2% of our DNA sequence, explaining both our close evolutionary relationship and the molecular basis for the differences that make us distinctly human.

The Sequencing Revolution Continues

The completion of the human genome was not an ending but a beginning. The same sequencing technologies that had made the Human Genome Project possible continued to improve at an exponential pace. In the mid-2000s, a new generation of sequencing technologies called "next-generation sequencing" or "massively parallel sequencing" emerged, using completely different approaches from Sanger sequencing. These methods could sequence millions of DNA fragments simultaneously rather than one at a time, increasing throughput by orders of magnitude while dramatically reducing costs.

The cost of sequencing a human genome plummeted from three billion dollars in 2003 to one million dollars by 2007, to ten thousand dollars by 2011, to one thousand dollars by 2014. As of 2025, human genome sequencing costs a few hundred dollars and takes less than a day. What was once the work of an entire decade for an international consortium involving thousands of scientists can now be done in a single laboratory in less than twenty-four hours.

This democratization of sequencing has transformed biological and medical research. Cancer researchers routinely sequence tumors to identify the specific mutations driving each patient's cancer, enabling personalized treatment strategies. Evolutionary biologists sequence genomes from hundreds of species, reconstructing the tree of life and understanding how organisms have adapted to different environments. Microbiologists sequence the complex communities of bacteria living in soil, in oceans, and in the human gut, revealing ecosystems of staggering diversity. Ancient DNA researchers sequence DNA from fossils, revealing the genomes of extinct species like Neanderthals and woolly mammoths.

Personal genomics companies offer direct-to-consumer genetic testing, allowing anyone to sequence parts of their genome for a few hundred dollars to learn about ancestry, disease risks, and other genetic traits. Millions of people have taken advantage of these services, creating enormous databases of human genetic variation that researchers use to understand human diversity and disease.

The Medical Revolution

The ability to rapidly and cheaply sequence DNA has begun transforming medicine, though realizing the full promise of "genomic medicine" remains a work in progress. Newborn screening programs in some countries now include genome sequencing to identify genetic diseases early when treatment might be most effective. Cancer treatment increasingly involves sequencing tumors to identify actionable mutations that can be targeted with specific drugs. Rare disease diagnosis, which once might have taken years and involved invasive procedures, can sometimes be accomplished with a simple blood test and genome sequencing.

Pharmacogenomics understanding how genetic variation affects drug responses allows doctors to prescribe medications based on patients' genetic profiles, avoiding drugs likely to be ineffective or cause adverse reactions in particular individuals. Prenatal testing can now detect genetic abnormalities in fetuses using DNA circulating in the mother's blood, providing information about fetal health without the risks of invasive procedures like amniocentesis.

Yet challenges remain. Having a genome sequence doesn't automatically translate into medical benefit. Interpreting sequence variations determining which are harmless and which cause disease requires extensive research and careful clinical validation. The human genome is full of variants of unknown significance, creating uncertainty rather than clarity. Privacy concerns about genetic information, potential discrimination based on genetic risk, and equitable access to genomic technologies all present ethical and practical challenges that society is still working to address.

The ability to read DNA to sequence genomes quickly and cheaply has moved humanity from a time when genes were mysterious and invisible to a time when we can read our genetic instruction book as easily as we read text on a page. We can see precisely how genes differ between individuals, how mutations cause disease, and how evolution has shaped the diversity of life on Earth. The power to read DNA has opened doors to understanding biology at a level of detail that would have seemed like science fiction just decades ago, and it has set the stage for the next revolution: not just reading DNA, but writing and editing it.

Fifteen

Manipulating DNA

"The Genetic Engineering Revolution"

Reading DNA sequences was revolutionary, but it was only half of the equation. To truly understand how genes worked, to develop new medicines, to improve crops, or to cure genetic diseases, scientists needed not just to read DNA but to manipulate it cut it, paste it, move pieces around, and ultimately to write new genetic instructions. The development of genetic engineering tools in the 1970s and 1980s gave scientists these capabilities, transforming molecular biology from a descriptive science into an interventional one and launching the biotechnology industry that would eventually be worth trillions of dollars.

The foundation of genetic engineering was laid in the late 1960s and early 1970s with the discovery of restriction enzymes, molecular scissors that could cut DNA at specific sequences. These enzymes had evolved in bacteria as defense mechanisms against viral infection bacteria used them to cut up viral DNA while protecting their own DNA through chemical modifications. Different restriction enzymes recognized different short sequences and cut

DNA at or near those sequences with exquisite specificity.

Werner Arber in Switzerland, Hamilton Smith at Johns Hopkins University, and Daniel Nathans at Johns Hopkins (who applied restriction enzymes to mapping and analyzing viral genomes) shared the 1978 Nobel Prize in Physiology or Medicine for their work on restriction enzymes. Their discoveries provided molecular biologists with a precise toolkit for cutting DNA at defined locations, essential for any kind of genetic manipulation.

Recombinant DNA

The next crucial development was recombinant DNA technology the ability to combine DNA from different sources into single molecules. Paul Berg at Stanford University pioneered this work in the early 1970s, developing methods to cut DNA from different organisms with restriction enzymes, then join the pieces together using DNA ligase enzymes that could seal breaks in DNA's sugar-phosphate backbone. Berg created the first recombinant DNA molecules in 1972, combining DNA from a monkey virus with DNA from a bacteriophage.

The implications were staggering. If DNA from any organism could be cut with restriction enzymes and joined to DNA from any other organism, then genes could be moved between species. A human gene could be inserted into bacteria, yeast, or other organisms. Genes could be modified, improved, or combined in new ways. The toolkit of genetic engineering suddenly made possible experiments and applications that had previously existed only in imagination.

But the power of recombinant DNA technology also raised concerns. What if dangerous combinations were created accidentally? What if genetically modified bacteria escaped from laboratories and caused environmental or health problems? In 1975, scientists gathered at the Asilomar Conference Center in California to discuss the potential risks of recombinant DNA research and establish safety guidelines. The Asilomar conference established important precedents for scientific self-regulation and public discussion of emerging technologies, though debates about the risks and benefits of genetic engineering continue to this day.

The First GMOs

The first practical application of genetic engineering came in 1978 when scientists at Genentech, a startup biotechnology company in San Francisco, successfully inserted the human gene for insulin into bacteria and got the bacteria to produce human insulin. This was a breakthrough for multiple reasons. It demonstrated that genetic engineering could solve practical problems diabetics needed insulin, and the traditional source (pig and cow pancreases from slaughterhouses) was limited and sometimes caused immune reactions. It proved that bacteria could be used as factories to produce human proteins. And it showed that biotechnology could be a profitable business, attracting investment and spawning an entire industry.

The insulin achievement required solving numerous technical challenges. The human insulin gene had to be identified and isolated. It had to be inserted into bacteria in a form they could express bacteria don't have the same gene regulation machinery as humans, so the gene had to be engineered with bacterial control sequences. The bacteria had to be grown in large quantities, and the insulin had to be purified from the bacterial culture. Each step presented obstacles that had to be overcome through clever molecular biology.

Genentech's success demonstrated the power and potential of genetic engineering, and it sparked a biotech boom. Companies formed to produce other medically important proteins: human growth hormone for treating growth disorders, clotting factors for hemophilia patients, tissue plasminogen activator for dissolving blood clots in stroke patients, and dozens of others. The era of biotech medicines had begun, eventually generating tens of thousands of jobs and hundreds of billions of dollars in economic value while providing treatments for previously untreatable diseases.

Agricultural biotechnology followed a similar trajectory. In the 1980s and 1990s, scientists developed methods to insert genes into plants, creating genetically modified crops with new characteristics. The first commercially grown GM crop was the Flavr Savr tomato in 1994, engineered to ripen more slowly and thus have longer shelf life. More practically important were crops engineered for pest resistance and herbicide tolerance. Bt corn contained a gene from the bacterium Bacillus thuringiensis that produced a protein toxic to certain insect pests, reducing the need for chemical pesticides. Roundup Ready soybeans were engineered to be resistant to the herbicide glyphosate, allowing farmers to control weeds more effectively.

By 2025, genetically modified crops are grown on hundreds of millions of acres globally, with traits including pest resistance, herbicide tolerance, drought tolerance, and enhanced nutrition. The technology has been controversial, with debates about environmental impacts, corporate control of agriculture, and food safety, but scientific evidence has generally supported the safety of approved GM crops while acknowledging legitimate concerns about agricultural sustainability and biodiversity.

Cloning

Perhaps no application of genetic engineering captured public imagination and generated more controversy than cloning creating genetically identical copies of organisms. The term "cloning" actually covers several different techniques, from cloning genes in bacteria (routine in molecular biology) to cloning entire organisms (much more difficult and controversial).

The breakthrough in organism cloning came in 1996 when Ian Wilmut and colleagues at the Roslin Institute in Scotland announced the birth of Dolly the sheep, the first mammal cloned from an adult somatic cell. Dolly was created by taking the nucleus from an adult sheep's mammary cell and inserting it into an egg cell whose nucleus had been removed. The reconstructed egg was stimulated to begin dividing and was implanted into a surrogate mother sheep, eventually developing into Dolly a genetic copy of the sheep that donated the adult cell.

Dolly's birth demonstrated that the genetic program of an adult specialized cell could be reset, returning to an embryonic state capable of directing complete development. This had profound implications for developmental biology, showing that cellular differentiation wasn't necessarily permanent and that adult cells retained all the genetic information needed to create an entire organism. It also raised immediate questions about whether humans could or should be cloned, generating ethical debates that continue today.

Since Dolly, numerous other species have been successfully cloned, including cattle, pigs, horses, cats, dogs, and even endangered species. Cloning has practical applications in agriculture (reproducing animals with desirable traits) and conservation (potentially rescuing species from extinction), though the technology remains inefficient and expensive. Reproductive cloning of humans remains prohibited by law and scientific consensus in virtually all countries, though therapeutic cloning creating cloned embryos as sources of stem cells for medical research continues in some jurisdictions under strict regulations.

Gene Therapy

If genes can be inserted into organisms to give them new traits, could defective genes causing human diseases be replaced with functional ones? This possibility, called gene therapy, has been pursued since the 1980s as a potential cure for genetic diseases. The concept is straightforward: identify the defective gene causing a disease, create a corrected version of that gene, and deliver it to the patient's cells using a virus or other vector that can insert DNA into cells.

The reality proved much more challenging than the concept. Early gene therapy trials in the 1990s encountered serious problems, including inefficient gene delivery, immune reactions to the viral vectors, and inadequate control over where inserted genes integrated into chromosomes. A devastating setback came in 1999 when Jesse Gelsinger, an 18-year-old participating in a gene therapy trial, died from a severe immune reaction to the viral vector. His death led to increased scrutiny of gene therapy research and a more cautious approach to clinical trials.

But gene therapy research persisted, and gradually successes began accumulating. In 2012, the European Union approved the first gene therapy treatment for a rare disease caused by lipoprotein lipase deficiency. In 2017, the FDA approved gene therapy for an inherited form of blindness and for certain types of cancer. As of 2025, multiple gene therapy treatments have been approved for various diseases, including some forms of inherited blindness, severe combined immunodeficiency, hemophilia, and sickle cell disease. The field has matured from promising concept to established medical treatment, though high costs and limited availability remain significant challenges.

CRISPR: The Gene Editing Revolution

The most trans formative recent development in genetic engineering is CRISPR-Cas9, a gene editing technology that allows precise, targeted modifications to DNA sequences in living cells. CRISPR (Clustered Regularly Interspaced Short Palindromic Repeats) is derived from a bacterial immune system that cuts foreign DNA. Jennifer Doudna at Berkeley and Emmanuelle Charpentier, then in Sweden, showed in 2012 that the CRISPR system could be programmed to cut DNA at any desired sequence, effectively creating customization molecular scissors.

CRISPR is revolutionary because it is far easier, faster, cheaper, and more precise than previous gene editing methods. Where older techniques might take months and cost tens of thousands of dollars to edit a single gene, CRISPR can accomplish the same task in weeks for hundreds of dollars. The technology has been rapidly adopted by research laboratories worldwide, accelerating biological research across nearly every field from basic cell biology to agriculture to medicine.

The applications are broad and growing. Researchers use CRISPR to study gene function by creating knockout mutations in model organisms. Agricultural scientists are developing CRISPR-edited crops with improved yields, better nutrition, or enhanced resilience to climate change. Medical researchers are exploring CRISPR-based treatments for genetic diseases, cancer, and infectious diseases. In 2020, Doudna and Charpentier were awarded the Nobel Prize in Chemistry for developing CRISPR-Cas9, barely eight years after their initial publication remarkably fast recognition reflecting the technology's tremendous impact.

But CRISPR also raises profound ethical questions, particularly regarding its potential application to human embryos. Editing genes in embryos would create heritable changes passed to future generations, effectively redesigning the human germline. In 2018, Chinese scientist He Jiankui announced that

he had created the first gene-edited babies, using CRISPR to modify embryos to be resistant to HIV infection. The announcement generated worldwide condemnation. He was criticized for proceeding without proper ethical oversight, for choosing an application that wasn't medically necessary, and for exposing the edited babies to unknown risks. He was subsequently sentenced to three years in prison by Chinese authorities.

The He Jiankui incident crystallized ongoing debates about the appropriate uses of gene editing. Should we edit human embryos at all? If so, for what purposes only to prevent serious diseases, or also to enhance traits like intelligence or physical abilities? Who should decide what edits are acceptable? How do we ensure equitable access to these technologies rather than creating genetic haves and have-nots? These questions have no easy answers, and societies are still working to establish appropriate regulations and ethical frameworks.

The power to manipulate DNA to move genes between organisms, to modify existing genes, to write entirely new genetic instructions has given humanity capabilities that would have seemed like magic just decades ago. We can produce medicines that were impossible to obtain from natural sources. We can improve crops to feed growing populations in changing climates. We can cure genetic diseases by correcting mutations at their source. But with these capabilities comes responsibility to use these tools wisely, to consider long-term consequences, to ensure benefits are distributed fairly, and to respect both the complexity of biological systems and the dignity of human life. The genetic engineering revolution is far from over; in many ways, it is just beginning.

Sixteen

DNA Today and Tomorrow

"The Continuing Story"

The journey from Mendel's monastery garden to CRISPR gene editing spans less than two centuries, yet in that time, humanity has progressed from complete ignorance about the molecular basis of heredity to the ability to read and rewrite the genetic code itself. This acceleration of knowledge and capability shows no signs of slowing. DNA technology continues to advance at breathtaking speed, opening new applications and raising new questions faster than society can fully absorb or regulate them. The story of DNA is not a closed chapter in the history of science but an ongoing narrative that touches virtually every aspect of modern life.

DNA Forensics: The Molecular Fingerprint

One of the earliest practical applications of DNA technology outside medicine and agriculture was forensic DNA profiling, often called DNA fingerprinting. The technique was developed in 1984 by Alec Jeffreys at the University of Leicester in England, who discovered that certain regions of human DNA contain short sequences repeated different numbers of times in different individuals. By analyzing these variable number tandem repeats (VNTRs) at multiple locations across the genome, Jeffreys could create a genetic profile that was effectively unique to each individual.

The first application came quickly. In 1986, Jeffreys used DNA profiling to exonerate one suspect and identify the actual perpetrator in a brutal rape and murder case in Leicestershire. The case demonstrated DNA's power for both excluding the innocent and identifying the guilty, and it launched DNA forensics as a standard tool in criminal investigations worldwide. Within a decade, DNA databases were being established in multiple countries, collecting genetic profiles from convicted criminals and crime scene evidence to help solve crimes.

DNA forensics has proven extraordinarily powerful. Cold cases decades old have been solved when DNA evidence preserved from crime scenes was analyzed with modern techniques and matched against DNA databases. Innocent people wrongly convicted have been exonerated when DNA testing excluded them as perpetrators of crimes they didn't commit. The Innocence Project, founded in 1992, has used DNA evidence to exonerate over 375 wrongly convicted individuals in the United States alone, including 21 who had been sentenced to death. These exoneration have exposed serious flaws in the criminal justice system while demonstrating DNA's reliability as evidence.

Modern forensic DNA analysis can work with tiny samples a single hair, a drop of blood, cells left behind when someone touches an object. The techniques have become so sensitive that investigators must be careful about contamination and about distinguishing between DNA deposited during a crime and DNA present for innocent reasons. Issues of privacy, consent, and the scope of DNA databases remain contentious, with ongoing debates about who should be included in databases, how long profiles should be retained, and what balance between public safety and privacy rights is appropriate.

Beyond criminal investigations, DNA identification has been crucial for identifying victims of mass disasters, wars, and terrorist attacks. When traditional identification methods fail because bodies are fragmented or burned, DNA testing of remains against reference samples from family members can provide definitive identifications, giving families closure and allowing proper burial. DNA testing also resolves paternity disputes, reunites families separated by war or disaster, and helps adopted individuals find biological relatives.

Personalized Medicine: Your Genome, Your Treatment

The falling cost of genome sequencing has made personalized or precision medicine increasingly practical. The concept is straightforward: different patients respond differently to the same treatments because of genetic differences, so tailoring treatments to individual patients' genetic profiles should improve outcomes while reducing adverse effects. The reality of implementing personalized medicine has proven more complex than the concept, but progress is accelerating.

Pharmacogenomics understanding how genetic variation affects drug metabolism and response has moved from research to clinical practice. Certain drugs are now routinely prescribed or dosed based on genetic tests. For example, the blood thinner warfarin requires careful dosing to avoid dangerous bleeding or ineffective clot prevention, and genetic variants in enzymes that metabolize warfarin significantly affect the appropriate dose.

Testing for these variants allows doctors to prescribe optimal starting doses. Similarly, certain cancer drugs are only effective in tumors with specific genetic mutations, and genetic testing of tumors guides treatment selection.

Cancer treatment has been particularly transformed by genomic approaches. Tumors from different patients with the same type of cancer say, lung cancer can have very different genetic mutations driving their growth. Comprehensive genomic profiling of tumors identifies the specific mutations present, allowing oncologists to select targeted therapies that attack those particular vulnerabilities. Some patients have dramatic responses to drugs that wouldn't have been tried before genomic profiling, while others are spared ineffective treatments and their side effects. The vision of truly personalized cancer treatment based on each tumor's genetic landscape is becoming reality.

Newborn screening, which has existed for decades to test for a handful of genetic diseases, is beginning to incorporate genomic sequencing. Some pilot programs now sequence newborns' genomes to screen for hundreds of genetic conditions simultaneously, potentially allowing early interventions that could prevent or minimize disease symptoms. This raises complex ethical questions about information management should parents be told about genetic risks that won't manifest until adulthood, if at all? How much information is too much? What psychological and social impacts might result from knowing one's genetic risks from birth?

Ancestry and Evolution: DNA as History Book

The same sequencing technologies that enable medical applications also allow reconstruction of human history and evolution with unprecedented precision. Companies offering direct-to-consumer genetic testing have collected DNA data from tens of millions of customers, creating enormous databases of human genetic variation. While marketed primarily for ancestry information and health risk assessment, these databases also serve as invaluable research resources for understanding human genetics, migration patterns, and population history.

DNA analysis has revolutionized our understanding of human evolution and prehistory. Sequencing DNA from Neanderthal fossils work that earned Svante Pääbo the 2022 Nobel Prize in Physiology or Medicine revealed that Neanderthals and modern humans interbred, and that most people of European and Asian descent carry 1-2% Neanderthal DNA in their genomes. This Neanderthal DNA isn't random some of it affects immune function, skin and hair characteristics, and even susceptibility to certain diseases. We carry within us genetic echoes of extinct human species, physical evidence of encounters between different human populations tens of thousands of years ago.

Ancient DNA from archaeological sites traces human migrations across continents, showing when different populations split apart or came back together. DNA from domesticated animals and crops reveals when and where domestication occurred. DNA from historical figures settles questions about relationships and identities. Richard III of England was identified from bones discovered under a parking lot partly through DNA matching to living descendants. DNA has become an essential tool for archaeologists, anthropologists, and historians, adding molecular evidence to supplement traditional historical sources.

Population genetics using modern DNA samples traces ancestry and relationships between populations, revealing the complex history of human migration, admixture, and adaptation. The DNA of different populations carries signatures of their histories bottlenecks when population sizes crashed, founder effects when small groups established new populations, selection signatures where genes conferring advantages in particular environments became more common. African populations show the greatest genetic diversity, consistent with Africa being the long-term home of human evolution before some populations migrated to other continents.

For individuals, ancestry testing can reveal unexpected family histories, connect people to distant relatives, and provide insights into geographic origins. These discoveries can be emotionally powerful, sometimes uncovering family secrets or challenging people's understanding of their identity. The social implications of widespread genetic ancestry information are still unfolding, with questions about how genetic ancestry relates to cultural identity, how information should be managed to protect privacy, and how to prevent misuse of genetic data for discrimination.

Synthetic Biology: Designing Life

If genetic engineering involves modifying existing genes and organisms, synthetic biology takes the more ambitious approach of designing and building entirely new biological systems. Synthetic biologists approach living cells as programmable systems that can be engineered like electronic circuits, with standardized genetic parts that can be combined in novel ways to create organisms with new capabilities.

Early synthetic biology milestones included creating simplified versions of natural genetic circuits, such as genetic toggle switches that could flip between two stable states, or oscillators that created rhythmic gene expression patterns. These demonstrations showed that biological systems could indeed be engineered with predictable behaviors. More ambitious projects have created bacteria with expanded genetic codes adding new base pairs beyond the natural A-T and G-C pairs or bacteria with entirely synthetic genomes designed and assembled in the laboratory.

The applications of synthetic biology span from mundane to trans formative. Engineered microorganisms produce medicines, biofuels, industrial chemicals, and specialty materials. Yeast have been engineered to produce artemisinin, an antimalarial drug that is difficult and expensive to extract from plants. Bacteria have been designed to detect environmental pollutants, to produce biodegradable plastics, or to manufacture flavors and fragrances. The potential is enormous, though realizing it requires overcoming numerous technical challenges around ensuring engineered organisms perform reliably and predictably.

More speculative applications venture into controversial territory. Could synthetic biology create organisms that photosynthesize more efficiently, helping address climate change by capturing more atmospheric carbon? Could we design organisms that live in extreme environments, perhaps enabling life support systems for space exploration? Could we resurrect extinct species by synthesizing their genomes and implanting them in cells of closely related living species? These possibilities raise profound questions about the extent to which humans should redesign the living world and what responsibilities we have to consider long-term and unintended consequences.

DE-extinction bringing back extinct species has moved from science fiction to serious discussion. Techniques for DE-extinction range from selective breeding (breeding living species to restore traits of extinct relatives), to cloning (using preserved DNA from extinct species), to genetic engineering (modifying living species' genomes to match extinct species). Projects are underway to restore extinct species including the woolly mammoth, the passenger pigeon, and the thylacine. Proponents argue DE-extinction could help restore damaged ecosystems or right past wrongs. Critics question whether resources would be better spent protecting currently endangered species and whether resurrected species could meaningfully fill their ecological roles in drastically changed environments.

The Data Deluge: Managing Genetic Information

The flood of genetic data being generated from medical sequencing, consumer testing, research studies, and agricultural applications has created new challenges around data management, privacy, and security. Genetic information is uniquely personal, unchangeable throughout life, and shared among relatives. This makes it both valuable and potentially problematic.

Medical applications require extensive databases linking genetic variants to health outcomes. The more data available, the better researchers can distinguish harmless variants from disease-causing ones and understand how genetic background modifies disease risk. Major initiatives like the UK Bio bank (500,000 participants), the NIH's All of Us program (aiming for one million participants), and similar efforts globally are collecting genetic data alongside detailed health information to enable research correlating genetics with disease.

But genetic data is also sensitive. Knowledge of genetic disease risks could affect individuals' self-perception and life choices, and could potentially be used for discrimination by insurers, employers, or others. Laws like the U.S. Genetic Information Nondiscrimination Act attempt to prevent such discrimination, but enforcement is imperfect and coverage is incomplete. Data breaches could expose genetic information. Direct-to-consumer testing companies hold genetic data on millions of customers, raising questions about data security, law enforcement access, and whether data might be sold or used in ways customers didn't anticipate.

The interconnected nature of genetic data means your genetic information reveals information about your relatives as well. If you submit your DNA to an ancestry database, you're potentially revealing information about siblings, parents, children, and more distant relatives who haven't consented. Law enforcement has used this feature to solve crimes by searching ancestry databases for relatives of suspects, a practice that has solved several high-profile cold cases but raises privacy concerns. Balancing the benefits of genetic data sharing against privacy rights remains an ongoing challenge with no easy solutions.

The Ethical Horizon

As DNA technology becomes more powerful, ethical questions multiply and intensify. Some issues that seemed theoretical just years ago are now practical realities demanding decisions. Human germ line editing making genetic changes that would be inherited by future generations—is technically feasible using CRISPR, but most countries prohibit or severely restrict it for humans. The 2018 scandal involving gene-edited babies born in China demonstrated both the feasibility and the widespread opposition to proceeding without proper ethical oversight.

Enhancement rather than treatment presents thornier ethical terrain. If we can edit genes to cure disease, can we edit genes to enhance normal traits increasing intelligence, athletic ability, or longevity? Who would have access to such enhancements? Would genetic enhancement exacerbate social inequalities? What would it mean for human nature and human dignity if we began consciously directing our own evolution? Different ethicists and different cultures reach different conclusions about where lines should be drawn.

Eugenics the idea of improving human populations through selective breeding or genetic intervention has a dark history, with horrific implementations in the twentieth century including forced sterilizations and genocides. Modern genetic technologies could enable forms of eugenics that are voluntary and individual rather than coercive and collective, but echoes of that history appropriately generate caution. How do we distinguish legitimate uses of genetic technology to prevent disease from problematic efforts to engineer "better" humans according to subjective criteria?

Environmental release of genetically modified organisms, particularly those designed to spread through populations using gene drives, raises concerns about unintended ecosystem impacts. Gene drives that could eliminate mosquitoes carrying malaria or modify agricultural pests have enormous potential benefits, but once released, such organisms might be impossible to contain or reverse. The precautionary principle suggests extreme caution; the humanitarian imperative to save lives suggests moving forward carefully but deliberately. Again, there are no easy answers, only difficult trade offs and the need for thoughtful governance.

Looking Forward

Standing in 2026, seven decades after the double helix structure was revealed, we can see how profoundly DNA has reshaped science, medicine, and society. But we may only be at the beginning of the DNA revolution. Current limitations imperfect gene editing, incomplete understanding of gene function, challenges delivering genetic therapies to the right cells are likely temporary obstacles that will be overcome with continued research and development.

The next decades will likely see routine genome sequencing for all newborns, with genetic information becoming a standard component of medical records used throughout life for disease prevention and treatment. Gene therapy and gene editing will cure genetic diseases that are currently fatal or disabling. Synthetic biology will provide sustainable alternatives to petroleum-based materials and chemicals. Precision agriculture guided by crop genomics will help feed a growing global population while reducing environmental impact.

More speculatively, we may develop capabilities to slow or reverse aging by understanding and addressing the molecular mechanisms of senescence. We may create designer organisms for applications we can barely imagine today living sensors, biological computers, custom organisms for terraforming other planets. We may gain the ability to preserve species facing extinction by maintaining their genetic information and potentially recreating them when conditions allow. The boundary between natural and designed life may blur as synthetic biology matures.

Yet with expanded capabilities comes expanded responsibility. The questions facing us are not just technical but moral, social, and political. How do we ensure that the benefits of DNA technology are distributed equitably rather than exacerbating inequalities? How do we respect human dignity and autonomy while enabling beneficial interventions? How do we balance innovation's potential benefits against risks of unintended consequences? Who gets to make decisions about technologies that could affect all of humanity and the biosphere we share?

These questions won't be answered by scientists alone. They require input from ethicists, policymakers, community representatives, and society as a whole. They require ongoing dialogue as technology advances and circumstances change. They require wisdom, humility, and a willingness to proceed carefully when consequences are uncertain. The power that DNA science has given us is immense, and wielding it responsibly may be humanity's greatest challenge and opportunity in the century ahead.

Seventeen

What DNA Teaches Us

"Lessons from the Double Helix"

The discovery of DNA's structure and function stands as one of humanity's supreme intellectual achievements, comparable to understanding planetary motion, atomic structure, or the origin of species. But beyond the technical accomplishments and practical applications, the story of DNA teaches profound lessons about life, about science, and about humanity's place in the universe. These lessons resonate across disciplines and deserve reflection as we contemplate what we've learned and what it means.

The Unity of Life

Perhaps the most philosophically significant insight from DNA research is the fundamental unity of all life on Earth. Every living organism from the simplest bacterium to the largest whale, from microscopic yeast to towering redwood trees, from viruses that blur the boundary between life and non-life to human beings contemplating their own existence uses DNA as its genetic material. All use essentially the same genetic code, translating the four-letter language of nucleotides into the twenty-letter language of amino acids using the same dictionary. All employ similar mechanisms for DNA replication, transcription, and translation.

This universality isn't coincidental. It points powerfully toward a single origin for all life on Earth, a last universal common ancestor from which every living thing descended through billions of years of evolution. We are literally related to every organism we encounter distant cousins to bacteria, closer relatives to other primates, close kin to other mammals. The same molecule carries genetic information in all of us, testifying to our shared heritage.

When we sequence the genomes of different species, the relationships become quantitatively precise. Humans share approximately 99.9% of their DNA sequence with other humans, regardless of geographic ancestry, ethnicity, or any other social categories humans create. We share roughly 98.8% of our sequence with chimpanzees, our closest living relatives. We share about 90% with mice, 85% with dogs, 60% with fruit flies, and even 50% with bananas. These numbers reflect evolutionary relationships the more recently two species shared a common ancestor, the more similar their DNA remains.

The similarities extend beyond mere sequence statistics. Many genes perform similar functions across wildly different organisms. Genes controlling basic cellular processes like energy metabolism are remarkably conserved. Even genes controlling development HOX genes that specify body plans show striking similarities between flies and humans. A gene crucial for eye development in mice can function in fruit flies, despite 500 million years of separate evolution. This deep conservation of genetic mechanisms reveals that life found solutions that work and preserved them across evolutionary time.

This unity has practical implications. Model organisms bacteria, yeast, fruit flies, mice, and others serve as stand-ins for studying human biology because the fundamental mechanisms are similar enough that discoveries in these organisms often apply to humans. Much of what we know about human genetics was first learned in organisms that are far easier and faster to study than humans. The universality of life's molecular machinery makes this transnational research possible.

But the unity of life also carries moral weight. Recognition of our biological kinship with other organisms arguably should inform how we treat the living world. Every organism we encounter, every species we might drive to extinction, shares with us a molecular heritage stretching back billions of years. We are not separate from nature but part of it, constructed from the same molecular building blocks, reading from the same genetic language.

Complexity from Simplicity

DNA teaches that enormous complexity can emerge from remarkably simple rules and components. The genetic code uses just four bases to encode all biological information. These bases pair according to simple rules—A with T, G with C. DNA polymerase follows straightforward chemistry to replicate DNA. Ribosomes translate mRNA using a systematic process. None of these individual steps is particularly complex.

Yet from these simple foundations emerges the stunning complexity of life. A human being develops from a single fertilized egg cell following instructions encoded in DNA. The genome orchestrates the development of trillions of cells of hundreds of different types, organized into tissues, organs, and systems that function together. The brain, containing billions of neurons with trillions of connections, develops according to genetic instructions combined with environmental influences. Behavior, personality, consciousness itself somehow emerge from molecules following chemical rules.

This progression from molecular simplicity to organismal complexity illustrates a fundamental principle: complex systems can arise from simple components following simple rules, given sufficient iterations and time. Evolution has had nearly four billion years to explore possibilities, to build and refine mechanisms, to create sophistication from simplicity. The result is living systems of breathtaking intricacy and elegance.

The lesson applies beyond biology. Complex behaviors emerge from simple rules in many systems weather patterns from physical laws, crystal structures from atomic interactions, intelligence from neural connections. The universe appears to be constructed such that complexity naturally emerges given the right conditions and sufficient time. DNA provides perhaps the clearest example of this principle, showing how four bases and straightforward pairing rules create the foundation for all biological complexity.

Yet DNA also teaches humility. Despite decades of intensive study, we still don't fully understand how genomes specify organisms. We can read genome sequences, but we can't always predict from sequence what traits will result. Genes interact with other genes in complex networks. Environmental factors influence gene expression. Development involves intricate feedback loops and emergent properties that resist simple reductionist explanation. The path from DNA to organism remains partially mysterious despite all our advances.

The Role of Information

DNA established that life fundamentally operates on information processing. Genetic information is stored in DNA sequences, copied during replication, expressed through transcription and translation, and ultimately manifests as the structure and function of organisms. This framing of biology in informational terms has been remarkably productive, connecting biology with information theory, computer science, and cybernetics.

The comparison between genetic information and computer code is both illuminating and limited. Like computer programs, genetic instructions specify processes that produce complex outcomes from simple rules. Both can be read, copied, modified, and debugged. Mutations in DNA are analogous to errors in code some harmless, some catastrophic, and rare ones beneficial. The modularity of genes resembles the modularity of subroutines in programming.

But organisms aren't just following a predetermined program. The relationship between genotype and phenotype between genetic information and actual traits is far more complex than the relationship between code and computation. Gene expression is context-dependent, influenced by environment, development stage, and previous history. Organisms actively modify themselves in response to conditions, showing plasticity that programs generally lack. And while programs are designed by programmers, genomes are products of evolution refined through countless iterations of variation and selection rather than consciously designed.

Nevertheless, thinking about genes as information has transformed biology. It enabled the development of bio informatics, applying computational methods to biological data. It provided frameworks for understanding regulation and control in biological systems. It suggested new therapeutic approaches, treating genetic diseases as information corruption that could be corrected. The information perspective has been one of molecular biology's most fruitful conceptual innovations.

The Interplay of Heredity and Environment

DNA research definitively showed that genes matter that inherited genetic variation influences traits and that mutations cause diseases. But it equally clearly showed that genes aren't everything. The old debate about nature versus nurture has been resolved: the answer is both, intertwined in complex ways that resist simple separation.

Even traits strongly influenced by genetics depend on environment for their expression. A gene that causes a disease might do so only in certain environments or with certain lifestyle factors. Identical twins, sharing 100% of their DNA, aren't completely identical in their traits or their disease susceptibilities, demonstrating that factors beyond DNA sequence matter. Epigenetics chemical modifications to DNA or associated proteins that affect gene expression without changing sequence provides mechanisms through which environment influences genetic expression.

The insight that genes and environment interact rather than operate independently has important implications. It means that genetic susceptibility to disease doesn't equal inevitability lifestyle interventions, environmental changes, or medical treatments may prevent or mitigate genetically influenced conditions. It means that genetic differences between individuals don't rigidly determine outcomes development, education, and life circumstances also profoundly matter. It means that arguments attributing complex human traits like intelligence, personality, or social outcomes solely to genes or solely to environment are both wrong.

This understanding should inform how we think about human diversity and social policy. Genetic differences between individuals exist and contribute to different traits and susceptibilities, but these differences occur against a background of much greater similarity, and their effects are mediated by environment, culture, and individual choices. Respect for human dignity and equality doesn't require believing everyone is genetically identical; it requires recognizing that genetic differences don't determine worth or potential, and that creating environments where all individuals can flourish is both possible and essential.

The Power and Limits of Reductionism

The success of molecular biology represents a triumph of reductionism the approach of understanding complex systems by studying their components. By breaking down heredity to chromosomes, genes, DNA, and ultimately base sequences, scientists achieved precise understanding impossible at higher levels of organization. The double helix structure, the genetic code, the mechanisms of replication and transcription all emerged from reductionist investigation.

Yet the DNA story also reveals reductionism's limitations. Knowing a genome sequence doesn't automatically tell you what an organism looks like or how it behaves. The whole is often more than the sum of its parts, with emergent properties that can't be predicted from components alone. Development, physiology, behavior, and evolution all involve systems-level phenomena that require holistic as well as reductionist approaches.

Modern biology increasingly recognizes the need for both reductionist and integrative approaches. Systems biology attempts to understand how genes and other components interact in networks rather than studying them in isolation. Evolutionary developmental biology ("evo-devo") examines how genetic changes translate into morphological evolution through developmental processes. These approaches build on the reductionist foundation that molecular biology provided but supplement it with attention to higher-level organization and dynamics.

The lesson is that reductionism is a powerful tool but not a complete worldview. Understanding components is necessary but not sufficient for understanding complex systems. Biology needs both the precise molecular knowledge that reductionism provides and the integrative perspectives that recognize organization and context. The great challenge is combining these approaches productively rather than seeing them as competing alternatives.

Science as a Human Enterprise

Finally, the DNA story teaches important lessons about how science actually works. The traditional image of science as a purely rational, objective process where hypotheses are tested and theories accepted or rejected based solely on evidence doesn't quite match the messy reality. The discovery of DNA involved brilliant insights but also lucky breaks, fierce competition, questionable ethics, and biased assumptions that delayed progress.

The long delay in recognizing DNA's importance Miescher's 1869 discovery, Avery's 1944 transformation experiments, resistance to abandoning the protein hypothesis shows how preconceptions can blind scientists to evidence. The questionable ethics around Franklin's data and the subsequent controversies about credit show science's human side, with personal relationships, ambitions, and biases influencing outcomes. The race to sequence the human genome shows how competition can drive progress but also create tensions between open science and proprietary interests.

Yet the story also shows science's self-correcting nature. Wrong ideas were eventually overturned by accumulating evidence. Overlooked contributors are gradually being recognized. Ethical lapses generate discussions that improve future practices. Science progresses not through individual genius working in isolation but through collaborative effort spanning generations, with countless scientists contributing pieces to an ever-growing puzzle.

Understanding this human dimension of science should make us appropriately humble about current knowledge and appropriately cautious about claims to certainty. Scientific understanding is always provisional, always subject to revision in light of new evidence. The greatest scientists have made errors; the most well-established theories have been overturned. We should hold current understanding firmly enough to base decisions on it but lightly enough to revise when evidence demands.

At the same time, science's human dimension shouldn't undermine trust in scientific knowledge. The fact that scientists sometimes make mistakes, have biases, or compete for credit doesn't mean scientific findings are arbitrary or unreliable. The process, despite its messiness, has produced genuine understanding of the natural world, including the profound insights about life and heredity that DNA research has revealed. The lesson is not that science is fatally flawed but that it's a human enterprise requiring continued attention to ethics, equity, and honesty.

Eighteen

The Future of Life

"Writing Tomorrow's Story"

As we stand at this moment in history, with the power to read and edit the genetic code, we face both extraordinary opportunities and profound responsibilities. The scientific journey that began with Mendel counting peas and culminated in CRISPR gene editing has given humanity capabilities unprecedented in the four billion years that life has existed on Earth. For the first time, one species our species can consciously direct its own genetic future and that of other organisms. How we exercise this power will shape the trajectory of life for centuries or millennia to come.

The Medical Frontier

The most immediate and least controversial applications of DNA technology lie in medicine. Genetic diseases that have caused immense suffering throughout human history may finally be curable. Sickle cell disease, a painful and life-threatening condition caused by a single mutation in the gene for hemoglobin, has been successfully treated using CRISPR to edit patients' blood stem cells. Several patients who underwent this treatment have been cured, freed from the disease that had dominated their lives. Similar approaches are being developed for beta-thalassemia, muscular dystrophy, and other genetic disorders.

Cancer, fundamentally a disease of DNA where accumulated mutations cause cells to grow out of control, is increasingly treated as a genetic disease requiring genetic solutions. CAR-T cell therapy, which genetically engineers patients' immune cells to recognize and attack cancer cells, has achieved remarkable success treating certain leukemias and lymphomas. As we better understand the genetic landscape of different cancers, more targeted therapies become possible. The vision of personalized cancer treatment, where each patient's tumor is sequenced and treatment tailored to its specific mutations, is becoming reality in major cancer centers.

Rare diseases, often caused by mutations in single genes, collectively affect millions of people worldwide. Individually rare, collectively common, these diseases have historically been neglected because each affects too few patients to make traditional drug development economically viable. But gene therapy and gene editing potentially offer generic platforms that can be adapted to different diseases by targeting different genes. As these technologies mature and costs decrease, treatments for rare diseases become more feasible.

Infectious diseases may also be combated through DNA technology. CRISPR-based diagnostics can detect pathogens rapidly and accurately. Genetically modified mosquitoes could eliminate malaria transmission. mRNA vaccines, which proved their worth during the COVID-19 pandemic, can be rapidly designed and manufactured for emerging infectious threats. The toolkit that DNA science has provided gives us unprecedented capabilities to respond to infectious disease challenges.

Yet medical applications raise difficult questions. Gene therapies currently cost hundreds of thousands to millions of dollars per patient, raising issues of access and equity. Should society guarantee access to life-saving genetic treatments, and if so, how should they be funded? How do we ensure that genetic medicine benefits all populations, not just wealthy ones? Clinical trials for genetic therapies are expensive and complex how do we balance the need for careful safety testing against the urgency felt by patients with currently unrepeatable diseases?

The Enhancement Dilemma

Beyond treating disease lies the more controversial territory of enhancement using genetic technology to improve normal traits rather than fix abnormal ones. The line between treatment and enhancement isn't always clear. Correcting a growth hormone deficiency is treatment; giving growth hormone to a short but normal child might be enhancement. Editing genes to prevent Alzheimer's disease is treatment; editing genes to boost normal intelligence would be enhancement. Where exactly should we draw lines, and who should draw them?

Some enhancements may be relatively uncontroversial. Few people would object to genetic interventions that extended healthy lifespan, keeping people active and independent longer before succumbing to age-related decline. Improving resistance to infectious diseases, enhancing DNA repair mechanisms to reduce cancer risk, or strengthening bones to prevent osteoporosis might be welcomed even if they go beyond treating current disease.

Other potential enhancements raise deeper concerns. Genetic modifications to increase intelligence, athletic ability, or physical attractiveness could exacerbate social inequalities if available only to the wealthy. They might reduce human diversity by pushing toward narrow ideals of what traits are desirable. They could put pressure on parents to enhance their children to keep up with others, turning procreation into an arms race. And they raise fundamental questions about human nature and authenticity would we remain recognizably human if we began consciously redesigning ourselves?

The most profound concerns arise around germ line editing making genetic changes that would be inherited by future generations. Somatic editing, which affects only the individual receiving the treatment, raises fewer ethical concerns because impacts are limited to consenting patients. But germ line changes would affect descendants who cannot consent, would spread through human populations over generations, and would be extremely difficult or impossible to reverse if problems emerged. Most countries currently prohibit germ line editing in humans, but pressure to reconsider may grow as technologies improve and potential benefits become clearer.

Environmental and Agricultural Applications

DNA technology promises to help address pressing environmental and agricultural challenges. Crops edited for drought tolerance, pest resistance, or enhanced nutrition could help feed growing populations while reducing environmental impact. Biofuels from engineered microorganisms could provide renewable alternatives to fossil fuels. Bacteria designed to break down plastics or other pollutants could help clean up environmental contamination. Trees modified to grow faster and store more carbon could help combat climate change.

Gene drives genetic modifications designed to spread rapidly through wild populations could eliminate disease vectors or invasive species. Mosquitoes carrying malaria could be driven to extinction or replaced with varieties that cannot transmit the parasite, potentially saving hundreds of thousands of lives annually. Invasive rodents devastating island ecosystems could be eliminated, allowing native species to recover. Agricultural pests could be controlled without pesticides.

But environmental applications also carry risks. Genetically modified organisms, once released, cannot be recalled. Ecological systems are complex and interconnected; changes intended to solve one problem might create others. Gene drives designed to eliminate one species might affect others unexpectedly. Engineered organisms might out compete wild relatives or transfer engineered genes to them through hybridization. The precautionary principle suggests moving carefully when consequences are potentially irreversible and far-reaching.

Governance of environmental genetic technologies remains inadequate. Who has authority to approve releases that might affect entire ecosystems or even the global environment? How should benefits and risks be weighed when different communities, nations, or generations might bear them differently? What processes should be followed to ensure decisions reflect diverse perspectives and values rather than narrow interests? These questions have no easy answers but demand attention before widespread environmental release of genetically modified organisms proceeds.

Preserving Biodiversity

Even as we gain power to modify organisms, we face the reality of accelerating extinction. Human activities habitat destruction, climate change, pollution, over exploitation are driving species to extinction at rates perhaps a thousand times higher than natural background rates. We are in the midst of Earth's sixth mass extinction event, and unlike the previous five, this one is caused by a single species us.

DNA technology offers some tools to address biodiversity loss. Genetic rescue involves introducing new genetic diversity into small, inbred populations to improve their health and adaptability. Genetic monitoring uses DNA to track wildlife populations, identify threats, and assess conservation strategies. Cryopreservation of genetic material creates "frozen zoos" that preserve species' genetic information even if living populations disappear, potentially allowing future recreation through cloning or synthetic biology.

DE-extinction recreating extinct species captures imaginations but raises complex questions. Should we invest resources in recreating woolly mammoths or passenger pigeons when thousands of currently living species need protection? Would recreated species be authentic or merely approximations with uncertain ecological roles? Might DE-extinction create moral hazard, suggesting species loss is reversible and thus reducing urgency for conservation?

The reality is that DNA technology, however powerful, cannot substitute for protecting habitats and ecosystems. It can be a tool supporting conservation but not a replacement for preventing destruction. The genetic information preserved in organisms is valuable, but genes alone don't constitute a species ecological relationships, behaviors, and adaptations matter too. Preventing extinction is far preferable to trying to reverse it after the fact.

The Wisdom to Choose

The trajectory of DNA science has been one of accelerating capability. Each decade brings tools and applications that seemed impossible or fantastic in the previous decade. This acceleration shows no signs of stopping; if anything, it's accelerating further. Within the lifetimes of people alive today, we will likely see genetic technologies that make current capabilities seem primitive.

The crucial question is not what we can do but what we should do. Technical feasibility doesn't imply ethical acceptability or societal desirability. As our powers expand, the need for wisdom, foresight, and inclusive decision-making grows more urgent. We need frameworks for thinking about long-term consequences, for weighing benefits against risks, for ensuring that the fruits of genetic technology are distributed justly rather than exacerbating inequality.

We need honest dialogue across boundaries between scientists and citizens, between experts and communities, between developed and developing nations, between current and future generations. We need to acknowledge that different people, cultures, and traditions may reach different conclusions about appropriate uses of genetic technology, and we need governance structures that respect this diversity while establishing guardrails against potential harms.

Most fundamentally, we need humility. The complexity of biological systems means unintended consequences are virtually certain. The interconnectedness of ecological and social systems means impacts will extend beyond narrow applications. Our knowledge, however impressive, remains incomplete and will always be provisional. History teaches that technologies often have unforeseen consequences, and the powerful technologies of genetic engineering are unlikely to be exceptions.

This humility shouldn't paralyze us with fear of proceeding, but it should inform how we proceed carefully, transparently, reversibly where possible, with attention to equity and justice, with willingness to pause or change course if problems emerge. The great biologist J. B. S. Haldane wrote, "Now, my own suspicion is that the universe is not only queerer than we suppose, but queerer than we can suppose." The living world we're learning to modify is certainly more complex and surprising than we fully understand. Proceeding with appropriate caution while embracing beneficial applications seems the wise middle path between paralysis and recklessness.

The Ultimate Question

Behind all the specific applications and ethical dilemmas lies a deeper philosophical question: what is humanity's proper relationship with nature? For most of human history, we saw ourselves as separate from nature perhaps above it, perhaps within it but distinct. Darwin showed we are products of nature, evolved like all organisms through natural processes. DNA research confirmed this at the molecular level we are made of the same stuff, running on the same genetic code, as bacteria and butterflies.

Yet we are also different. We are the first species that can understand its own genetic programming, can consciously modify it, can direct evolution deliberately rather than waiting for natural selection's slow winnowing. We are nature becoming conscious of itself, matter organized in such a way as to contemplate its own organization. This grants us both incredible power and profound responsibility.

Do we have the right to modify other organisms for our benefit? To modify ourselves? To redirect evolution? Or do we have an obligation to preserve nature as it is, to recognize limits to our interventions? These questions don't have simple answers because they involve values and worldviews, not just facts. Different philosophical and religious traditions, different cultural perspectives, different individual convictions lead to different conclusions.

Perhaps the wisest approach is to proceed with both boldness and humility boldly addressing suffering and solving problems where genetic technology offers solutions, but humbly recognizing the complexity we're dealing with and the limits of our understanding. To use our knowledge to reduce disease, preserve species, and improve human welfare, while respecting the intrinsic value of the natural world and the dignity of human life. To embrace the possibilities that DNA science offers while remaining mindful of its limitations and risks.

The story of DNA is ultimately a story about knowledge and power how we gained unprecedented understanding of life's molecular foundations, and what we will do with the powers that understanding provides. The next chapters of this story are being written now, not just by scientists in laboratories but by all of us through the choices we make, the policies we support, and the values we uphold. The discovery of DNA gave us the ability to read life's instruction manual. How we use that ability whether we prove wise stewards of the genetic knowledge we've gained will determine what kind of future we create for ourselves and for all life on Earth.

Nineteen

EPILOGUE: The Molecule That Changed Everything

There is something profoundly moving about tracing the arc of discovery from Mendel's monastery garden to the present day. In less than two centuries, humanity has progressed from complete ignorance about heredity's molecular basis to the ability to read, analyze, and edit the genetic code with precision. We've gone from vague speculation about invisible "factors" controlling inheritance to detailed understanding of DNA's structure and function. We've moved from passive observation of heredity to active manipulation of genes. The pace of this transformation is breathtaking, the scope of its implications almost impossible to fully grasp.

Consider what has been achieved. We understand how genetic information is stored in the sequence of nucleotide along DNA's sugar-phosphate backbone. We know how that information is copied during cell division, ensuring each daughter cell receives the same genetic instructions. We comprehend how DNA's message is read out through transcription and translation, directing the synthesis of proteins that perform cellular functions. We've deciphered the genetic code that translates nucleotide sequences into amino acid sequences. We can sequence entire genomes, reading billions of base pairs of genetic information. We can edit genes with precision, correcting mutations or introducing deliberate changes.

These capabilities have spawned entire industries. Biotechnology companies produce medicines insulin, growth hormone, clotting factors, antibodies using genetically modified organisms as factories. Agricultural companies develop genetically modified crops with enhanced traits. Forensic laboratories use DNA analysis to solve crimes and identify victims. Direct-to-consumer genetic testing companies offer ancestry analysis and health risk assessment to millions of customers. Gene therapy companies develop treatments for previously incurable genetic diseases. The economic impact runs into trillions of dollars; the medical and social impacts are literally incalculable.

Yet for all this progress, mysteries remain. We still don't fully understand how genomes encode organisms how the linear string of bases in DNA gives rise to three-dimensional form and function. The relationship between genotype and phenotype involves complexities we're only beginning to appreciate. Epigenetic modifications, regulatory networks, developmental dynamics, environmental interactions all add layers of complexity beyond the DNA sequence itself. We can read the genome, but reading isn't the same as understanding. We remain, in some profound sense, in the foothills of understanding even as we develop powerful applications based on partial knowledge.

The human dimensions of the DNA story deserve remembering alongside the scientific achievements. Brilliant scientists made crucial discoveries, but so did patient researchers whose contributions went unrecognized. Men dominated the story, but women made essential contributions, often while facing discrimination and sometimes being denied proper credit. Some scientists behaved ethically; others took shortcuts or appropriated colleagues' data. Competition drove rapid progress but also generated conflicts and casualties. The DNA story reminds us that science is a human endeavor, with all the nobility and flaws that implies.

The ethical challenges posed by DNA technology will intensify as capabilities expand. Should we edit human embryos to prevent disease? To enhance traits? Should we resurrect extinct species? Create entirely new organisms? Drive mosquitoes to extinction to eliminate malaria? Modify ourselves to slow aging or enhance intelligence? These aren't theoretical questions but practical dilemmas requiring decisions soon if not already. The technical capability to do something doesn't answer whether we should do it. Science can inform these choices but cannot alone resolve them; we need input from ethics, philosophy, religion, social science, and democratic deliberation.

Perhaps the most remarkable aspect of the DNA story is how completely it has re framed our understanding of life. Before DNA, heredity was mysterious, variation seemed random, and the relationships between organisms remained unclear. After DNA, we understand heredity as molecular information transfer, variation as the result of mutations in DNA sequences, and relationships between organisms as reflections of shared evolutionary history written in their genomes. The molecular unity of life all organisms using DNA and the same genetic code demonstrates kinship spanning the entire tree of life.

This molecular understanding hasn't diminished wonder at life's complexity and beauty; if anything, it has enhanced it. Knowing how DNA encodes proteins that self-assemble into intricate molecular machines, knowing how regulatory networks control development, knowing how evolution explores genetic possibility space through mutation and selection this knowledge reveals sophistication and elegance beyond anything imagined before molecular biology. The more we learn about how life works at the molecular level, the more amazing it appears that such complexity emerges from chemical interactions.

Looking forward, the DNA revolution shows no signs of ending. New sequencing technologies continue dropping costs while increasing speed and accuracy. CRISPR and other gene editing tools are being refined and improved. Synthetic biology advances toward being able to design organisms with novel capabilities. Artificial intelligence trained on vast genomic datasets is beginning to predict gene function and design proteins. The next decades will likely bring capabilities and applications we can barely imagine today, just as researchers in the 1950s couldn't have imagined genome sequencing or CRISPR.

The challenge facing us is to wield this knowledge wisely to use DNA science to reduce suffering, enhance human welfare, preserve biodiversity, and improve quality of life, while avoiding applications that would diminish human dignity, exacerbate inequality, or cause environmental harm. This requires ongoing dialogue, inclusive decision-making, thoughtful regulation, and willingness to say no to technically feasible applications that are ethically problematic. It requires balancing innovation against precaution, hope against humility, ambition against awareness of our limitations.

Standing at this moment in history, we can look back with appreciation at how far we've come from Mendel's pea plants to the human genome, from transformation in bacteria to CRISPR editing, from mysterious invisible factors to precise molecular understanding. And we can look forward with both excitement and trepidation at where we're headed toward capabilities to cure genetic diseases, extend healthy lifespan, enhance human capabilities, and modify organisms in ways previously impossible.

The discovery of DNA ranks among humanity's greatest intellectual achievements. The double helix structure revealed in 1953 opened doors to understanding that transformed biology from a descriptive science into a molecular one. The subsequent decades of research built a comprehensive picture of how genetic information is stored, transmitted, modified, and expressed. The development of technologies to read and manipulate DNA has given us unprecedented power over the living world.

What we do with this knowledge and power will define the next chapters of both human history and the history of life on Earth. Will we prove wise stewards of the genetic knowledge we've gained? Will we use DNA technology to create a better world healthier people, more sustainable agriculture, preserved biodiversity? Or will we unleash consequences we didn't foresee and cannot control? The answer depends on choices not yet made, by people not yet born, in circumstances we cannot fully predict.

What's certain is that the DNA story continues. Every day, in laboratories around the world, researchers are pushing the boundaries of what's possible, solving new puzzles, developing new applications. The molecule that Friedrich Miescher isolated from pus-soaked bandages in 1869, that Oswald Avery identified as the transforming principle in 1944, whose structure Watson and Crick revealed in 1953, remains at the center of biology and increasingly at the center of medicine, agriculture, and technology more broadly.

From that first moment when Mendel noticed patterns in his pea plants, a pattern suggesting discrete hereditary factors, through all the discoveries and all the applications that followed, to this present moment where we can read and rewrite the code of life itself this has been a journey of spectacular scientific achievement. But it has also been a human journey, full of brilliant insights and stubborn mistakes, spectacular successes and ethical lapses, recognition and neglect, hope and caution.

The molecule that changed everything DNA, the double helix, the stuff of genes continues to shape our world and our future. Understanding where we came from in this journey of discovery can help guide where we go next. The story of how we learned to read the book of life should inform how we decide what to write in its future pages. That story, which began in a monastery garden 160 years ago, is still being written today, and its next chapters will be among the most consequential in human history.

www.ingramcontent.com/pod-product-compliance
Lightning Source LLC
LaVergne TN
LVHW090407210926
848480LV00002B/482
* 9 7 9 8 2 3 3 0 9 6 4 8 8 *